Praise for *Green Home Buildin*

Miki and Doug nailed it! This book should be required reading for every architect, builder, and homeowner aiming to build a home. Green building is about how the house operates as a system. They explain the components, tradeoffs, and decisions that lead to an energy-efficient, comfortable, healthy, affordable home. Finally, a comprehensive book that spells out information I've been sharing with my clients for years.

— Wayne Jeansonne,
President, Solluna Builders LLC,
a custom homebuilder in Austin, Texas

Finally, a book about building green that's for the people. Miki Cook walked us through the construction of our very affordable four-star green energy home in Austin. With her guidance we learned that green construction is not about bells and whistles, nor is it a vanity project for those with means. It's about going back to the common-sense strategies of sustainability: selecting the right location, designing with the environment in mind, using only what you need, and choosing materials that are healthy for you and your surrounding community. Along with the Doug Barrett, she has written a very engaging book that covers all these principles and more.

— Eric Tang and Paula Rojas, Austin, Texas

In *Green Home Building*, Doug and Miki have provided an encyclopedic guide for home owners with ambitions to build their own green homes. Particular kudos are due for addressing some important but oft-neglected issues such as "shape factor" and the life of the building beyond today's needs of the household—what about 10, 30, 50 years hence? This will prove to be an invaluable resource for those committed to a deep understanding of what it means to build a green home.

— Ann V. Edminster, M.Arch., LEED AP,
author, *Energy Free: Homes for a Small Planet*

This book is dedicated to all
who have prepared this path and mentored us.

To all the great people in this industry
who continue to define *sustainability*.

And to all those who strive to do
their share for a better future for us all.

Thank you!

Disclaimer

Information included in this book related to green buildings, energy efficiency, water conservation, protecting our environment, healthy indoor air quality and cleaning practices and sustainable building practices are provided as general educational guidelines and are not meant to infer any specific guarantee of benefits. Information provided herein is not an exclusive representation of high-performance methods, materials or practices. Many other efficient, healthy, and low environmental-impact products and building systems are available in the market besides those referenced in this manual, and therefore errors and omissions may and probably do occur. Since variations of products, building materials and methods and other green systems discussed in this book may or may not be installed in your home, actual performance may vary.

Construction methods discussed in this book may require further investigation from sources outside of this book prior to implementation or determination for appropriate applications. The authors offer no warranty, express or implied, as to the completeness or appropriate use of these methods on any specific project or application. The authors and publisher do not assume any legal liability or responsibility for the accuracy, completeness or usefulness of any information, apparatus, product or process disclosed. The views and opinions of the authors expressed herein may be subject to change.

Be wary of products that claim sustainable or green contents or practices without providing substantiation, documentation or research. Only products that carry Green Seal, Greenguard, Energy Star, Forest Stewardship Council, Carpet and Rug Institute Green Label or other reputable third-party standard logos, or have referenced compliance against industry-recognized standards, have been tested and verified by independent verification methods and should be considered credible.

Contents

Foreword

By Sara Gutterman

Our world is unequivocally changing, perhaps faster and more irreparably than we realize. And we're finding that the planet is more sensitive than we thought. Remote places and creatures that we once believed would be the furthest removed from climate change are instead fighting for their lives on the front lines. Our built environment is being pummeled by extreme weather events, and communities across the globe are suffering consequences ranging from financial losses to complete displacement.

Data shows that we now have over 400 parts per million (PPM) of carbon dioxide in the atmosphere, and climate scientists have irrefutable proof that this level of carbon dioxide and other greenhouse gasses are causing rising temperatures, more frequent and extreme weather events, increased atmospheric moisture, and a pronounced "dome effect" in urban areas, trapping heat and pollution.

Climate change will impact lives and livelihoods in almost all areas throughout the 21st century: costal zones will face rising sea levels; urban areas will grapple with extreme heat and inland flooding; and rural areas will struggle with insufficient access to drinking water. As the world becomes hotter and more polluted, it is predicted that accessible fresh water will decline by a minimum of 20 percent, forests around the world—which are essential to carbon sequestration—will become decimated, and coral reefs will die en masse.

Throughout the 21st century, climate change impacts will slow economic growth, expand poverty in urban areas and emerging "hunger hotspots", and trigger new poverty pockets. Financial losses will increase exponentially with each degree rise in temperature.

While the facts can be daunting, we don't need to be afraid of the changes taking place. We just need to be realistic. This is the moment for climate action. The choices that we make today will determine the future of our planet, and all of the species on it.

We have the financial resources and proven technology to reach our carbon emissions and resource use targets, now all we need is the desire, determination, and discipline to make the right decisions. How will we respond in the face of adversity? Who will we become in these trying times? Will we rise to the challenge of building our cities and communities responsibly, sustainably, and environmentally appropriately? Will we develop the courage to deploy wisdom and long-term consideration that has been missing from our approach for too many decades?

When it comes to determining the biggest carbon offender, the verdict is unambiguous: the urban built environment, responsible for approximately 70% of global energy consumption and carbon emissions, is the main culprit.

Today, over half of the world's population lives in urban areas. It's predicted that cities will absorb an additional 1.35 billion people by 2030, accounting for nearly all of the projected global population growth in that timeframe.

Across the planet, there is approximately 1.6 trillion square feet of built space (75% residential and 25% commercial). It's expected that an additional 860 billion square feet of space will be newly constructed or retrofitted by 2030.

Given that a significant portion of the total built environment is expected to be built or rebuilt over the next two decades, those of us building or retrofitting homes have a singularly unique opportunity to implement solutions that will address our urgent environmental needs. The answer has become personalized, and it is now our imperative to create structures that are resource efficient, adaptable to our changing climate, and meet our emissions targets.

Green homes have never had as much opportunity as they do today to directly—and substantially—shape the future.

If we are successful in our attempts to mainstream green building and keep carbon emissions levels under 1 trillion tons, thereby maintaining an average temperature increase threshold of 2°C, we might just have a fighting chance. If not—if emissions and temperatures exceed these levels, experts predict that our climate system will spin out of control. Translation: game over for humans and many other species.

It's sobering and exciting to think that the future is so decisively in our hands—the same hands that design and build green homes around the world. It's our privilege to be the change that is needed in the world. It's our responsibility to care enough to do what is right.

In this light, I invite you to savor this book by Miki Cook and Doug Garrett. May this compilation of instructive content guide you in your quest to design and build greener structures so that you can be a part of the solution leading to a more sustainable future. Exponential change, after all, is the result of the incremental improvements made by individuals every day.

—Sara Gutterman
Chief Executive Officer
Green Builder Media
Building a Better World
www.greenbuildermedia.com

Introduction

According to the Department of Energy, residences account for 54 percent of all the energy[1] and 74 percent of water[2] used in buildings in the US, more than the commercial and industrial sectors combined. A single traditionally built wood-framed house can consume the lumber harvested from one acre of forest[3] and can send 2 to 11 tons of construction waste to the landfill.[4] These activities not only waste resources, they add costs without adding benefits. It's not difficult to see that, with a few simple changes in how we build homes, we can reduce resource depletion and costs at the same time.

Although some architects and builders have been developing and applying green practices for twenty to thirty years or more, the mainstream home building industry and general population of the United States have just begun the transition to more sustainable practices. And even though the green building movement has been getting a lot more press and attention over the last few years, the overwhelming majority of homebuilders still don't understand what it is or how to structure implementation in order to achieve real measurable benefits, including a worthwhile return on investment. Much of the information included in this book was developed to help this movement take root, providing a road map to achieving a truly green home within any budget.

The authors are both green building consultants with years of residential construction and building science background. Miki Cook spent many years in the design/build, purchasing and estimating cost fields working for one of the original builder members of the Austin Energy Green Building program, the oldest green building certification program in the US. Doug Garrett founded Austin Energy's Residential Energy Conservation programs. In 1996, Doug established the first building science consulting business in Texas, providing building science-based forensic investigations, diagnostics and design consultations for clients and homebuilders across the nation. Over their careers, both have witnessed far too many project goals abandoned due to budgetary constraints. This book was written to provide a different approach to green building, from start to finish, with proven strategies and methods to achieve all of your goals and stay on budget.

Whether you are a homeowner, architect or a homebuilder reading this book, we hope you engage the methods discussed to build (or remodel) affordable, high-performance, healthy homes. Even if you are buying an already-built home (or even a condo in a high-rise), by gaining an understanding of the strategies presented in this book, you will be better able to recognize the green potential of all the properties available in the market. You will also be able to analyze those properties for any opportunities to further improve their performance and benefits. This book is about making informed, educated decisions in order to achieve your long-term goals, and about understanding the synergy of how each decision affects everything else.

We're All Green with Envy!

Green is everywhere; everything is green. This is not far from the truth these days. It's difficult to pick up a magazine without the cover story providing some insight into how to green your lifestyle, or turn on the television without seeing an advertisement for some company's efforts to minimize the environmental impact of their manufacturing operations. For the most part, the truth is that almost all green products are really only some *shade* of green, depending on their embodied energies, the toxicity of their contents and the value of the benefits that they provide.

Green homes themselves come in shades of green and are likely to perform accordingly. Green building programs that provide verification of green-built homes vary widely in their mandatory requirements for certification or the à la carte credits offered to achieve levels of higher recognition. Two projects with the same level of certification could have chosen significantly different methods in achieving it. And those methods, if not implemented to provide synergistic performance improvements, may actually do little to provide measurable benefits.

Even production builders who build the same plans repeatedly throughout a neighborhood to the same set of specifications will see each home perform differently, based on its orientation and, of course, the occupant's lifestyle choices. It is fairly safe to say that every green home is different. In fact, in some cases, using the term "green" may only describe individual features in the home, not the home itself.

Many builders will claim their homes are green and may offer a variety of reasons. A group of builders from a rural area of the country was once heard promoting their twenty–year-old heritage as green builders, as one of them had recently read that a handful of their regional methods were considered green. Their lumber was harvested from within 500 miles, they installed kitchen and bathroom vents that exhausted 100 and 50 cubic feet per minute (CFM) respectively (industry product standards), and to save money, they had been buying concrete made with fly ash. It didn't matter that they did not install energy-efficient windows, use less toxic building materials and only offered carpet and vinyl flooring to their customers. Obviously, the term "green" is being stretched here.

What Is a Green Home?

In general, green is based on the concept of sustainability. The most commonly accepted definition of sustainability refers to our ability to meet our needs in the present without compromising the ability of future generations to meet their needs. The truth is that we are using up resources on this planet at an alarming rate, not sustainably—everything from fossil fuels to fresh water to forests and food resources—and in doing so, we are endangering the delicate balance of nature. We are impacting climate and ecosystems by overharvesting and polluting, leading to events that eventually may threaten our own survival. The effect of these practices has already devastated many important natural resources globally, including our rainforests, farmland and wildlife habitats.

The reality is that we have already begun to see the impact from overexploiting our natural resources on our health and economic well-being. With an exploding world population, it is easy to see that these issues may reach staggering proportions within our own lifetimes, some escalating at such an alarming rate we may see major impacts within only a few years. Green building is not just for future generations, it protects our own quality of life.

A truly green home should deliver real benefits in terms of the amount of energy and water it requires to operate over its lifetime. It should be built using durable materials and methods, so there is less maintenance and longer periods between major repair cycles. It should provide a comfortable shelter for its occupants, one with fewer environmental pollutants that affect their health. And it should place less of a burden on our community and our planet, not damaging ecosystems or requiring the creation of massive new infrastructure to support it.

A key concept employed throughout this book comes from a relatively new field of housing research called building science. Building science studies and views the house and all of its components as parts of an interactive, integrated, holistic system. The mantra of building science is: "the house *is* a system." Building science recognizes that changing one aspect of how a home is built changes the entire system, and often other aspects of the home must be changed in response. The big value added is that this can be done while improving your comfort, reducing maintenance headaches and costs and at the same time putting a lot of monthly utility dollars back into your pocket. It also recognizes that the right way to build is what is right for your particular climate zone, not some one-size-fits-all approach.

How Can I Build an Affordable Green Home?

Section One of this book provides an easy-to-follow outline of "Ten Steps to an Affordable, Healthy, High-Performance Home," approaching the various green features and strategies as to how they impact the cost of housing. In order to truly achieve housing affordability, we need to look not only at the initial construction budget, but also at other costs that can be attributed to our total cost over the lifetime of the home. We all anticipate that energy and water costs will continue to rise in the future, as will material costs for the maintenance and repair of homes over their lifespan. Additionally, a failure to address indoor air quality can lead to health issues that increase our medical expenses, and we've all seen the direction healthcare costs are going.

Also, where we build homes can have significant impact on not only our land costs, but also the costs associated with community services. Green developments have a lower impact on our community, the environment, ecosystems and wildlife, and thus lower our costs associated with those. Choosing the right location for your green home can greatly influence your total cost.

Green building promotes resource efficiency, but you should not assume that every green home is resource efficient. It just makes sense that the more resources you use, the more the home costs to build and operate. So, for an affordable green home, it is imperative that you don't use more resources than you need to, either during construction of your home or in its operations once you've moved in. Efficient use of resources means obtaining the maximum benefit from the least amount of resources. We will look at both the materials and how we use them to keep our green home affordable.

Green materials and systems can improve the energy efficiency and indoor air quality of our homes and, to some extent, help to conserve water. But oftentimes, the focus is solely on materials and systems, without recognizing that basic design improvements made early in the process could have achieved far superior efficiencies. This add-on approach to building green is typically what adds costs to projects and is the main reason that those homes fail to deliver the benefits that we expect. We will look at the building as a whole and how to achieve synergistic results.

This is not to say that high-performance systems are not a worthy piece of the puzzle. This book will discuss the benefits that mechanical systems can provide and how these systems should be incorporated into your project so that they provide true benefits. So, unless you believe that green buildings must encompass every new technology available on the market today, wherein "new technology" most often translates to more expensive, you can use the strategies discussed in this book to build a *real* green home with *real* green benefits affordably.

The fact is that much of what we now understand about building science has shown us that the old passive strategies that we used before we had expensive systems to keep us comfortable give us the edge that we need to take high-performance homes to the next level. These strategies are discussed throughout the book and provide the foundation for Section Two of this book, "Getting to Zero," taking us beyond the green-built home, exploring the opportunities to zero-out energy, water, resource depletion, greenhouse gases and costs. If you can incorporate these strategies into your project, they provide both initial and long-term benefits and address our uncertainties related to future energy, water and natural resource availability, environmental contamination, and housing affordability. Any discussion of green building would not be complete unless we consider that we are stewards of the ever-changing landscape of sustainability.

We're not saying that you cannot spend a lot of money building green. The truth is that you can spend a lot of money on any kind of building. If you choose expensive systems, fixtures and finishes, you will spend more money no matter what type of building design, systems, products or construction methods you use. Unless you integrate the design, construction materials, methods and systems into achieving synergistic performance, only time will tell if that added expense will deliver real benefits and if the return on investment (ROI) is justifiable. Our goal should be a return on investment for all cost premiums of no more than 15 years.

This book is not just a collection of green strategies that you can pick and choose from in order to build or remodel a green home. It is about the specific process that you can use to achieve a truly green home for the same total cost of ownership as a traditionally built home. This includes many green strategies that can be incorporated into your project that cost nothing. It includes green strategies that reduce the total cost related to your housing choices, including your property taxes and commute expenses. But mostly, this book is about green strategies that you can use to lower your base construction costs in order to fund recommended high-performance upgrades to lower maintenance and operation costs and protect your family's health. We really want you to learn about these; we want green building to be affordable. Because when it is affordable, its benefits are understood and you realize that you can actually make green building work for your project, then there will be no reason not to do it. And when you apply these methods to your project, and your neighbor applies them to his, and the same contractors are working for everyone and learn that these methods work well on every home, they will just become the way that we build all homes. This is our goal for this book: to bring affordable green building methods mainstream.

This does not mean that we are promoting cookie-cutter-style homes. On the contrary, we want to introduce an approach that can be used on any style home, and any size. These strategies can be applied to any budget, from low-income housing to luxurious high-end custom homes, in order to achieve a high-performance green home within that budget. Note that the models and case studies presented in this book present fairly modest dimensions and cost. As you will see early on in the text, we do promote methods and materials to save you money and right-size your home based on your needs. Some would question whether large homes can even be considered green. Our hope is that those who build large homes will still build them using methods that result in efficient use of resources and healthy environments.

Regardless of what price point of home you are planning to build, we hope that you incorporate as much as you can from this book into your plan. By approaching your project from this perspective, you should realize all your green goals within the same total cost of ownership over time as a traditionally built home. Hopefully you will be able to share with your friends and family ways that they might also be able to do this.

In many cases we will suggest tradeoffs, showing how you can save money in one area of construction and then use those savings to improve another area, at little or no net cost. We believe that green building must first be green from an economic perspective. For you and your family, it is critical that your new green home or renovation be affordable to build and cost effective to operate.

Green should not be beyond anyone's reach; it should not be what takes you over your budget. If we all accept responsibility for what we can do, we will make a huge, valuable contribution to the continued transformation of our society as a whole to living more sustainably on our fragile planet. Throughout this book, the strategies we describe will save you money either in initial construction, lifetime operations, or both. In addition, we've highlighted "No Cost Green" strategies or product selections that provide green benefits without any additional cost to your project. These are identified in the text by the icon below. And most importantly, look for the "Key" symbol. It identifies those strategies that are key to achieving major savings in your construction budget. These are the most critical overall to attaining your goal of an affordable green home.

No Cost Green

Key Strategy

Show Me the Money
Ten Steps to an Affordable, Healthy, High-Performance Home

CHAPTER 1

Location, Location, Location

For most people, selection of a building site is usually based on proximity to the area where they want to live, their budget and/or how much flexibility they have in building design. The location that you choose should match your needs for access to community amenities, work, good schools for your children, and/or your lifestyle goals. Too many times people make poorly thought-out decisions, such as building large, luxurious homes on acreage outside of town only to abandon them within a couple of years, realizing that they hate the commute to work every day.

Just as often, people buy undeveloped sites in established infill areas only to discover that these sites were never built on because of poor soils, drainage issues or development restrictions pushing development costs above neighborhood home values. Most people do not know enough about building to determine cost variances associated with site abnormalities. Sites with steep grade or poor soils may mean that engineering and installing the foundation will be cost prohibitive. If the site is heavily treed, it may be subject to city ordinances that prohibit the removal of large trees that just happen to be located within the allowable build area of the site. So it might be difficult to fit the home on the site or you may have to take expensive measures to prevent roots from damaging the foundation over the long term.

Some sites have expansive clay soils, which can cause foundation problems, while other sites are so rocky that they cannot support the vegetation that will be needed to shade the structure, much less any kind of landscaping. If your goal is to put a solar photovoltaic system on your home, some sites in dense urban areas might not have enough sun exposure, due to trees or proximity to neighboring structures, to support that feature. If possible, it is best to have your builder or architect review the site before you make an offer to buy to be sure it will accommodate the size and type of home you intend to build.

You should also research any future growth, development and planning studies available for your area. As communities and regions struggle with population migration issues, you'll want to know that your decision to build and live in those areas will be a worthwhile investment in the long run. A recently published study[1] indicates that, by the year 2050, 90 percent of the US population is expected to live in dense, urban areas (current 2010 census indicates we are already at 82 percent). There have been reports in the media that the expected volume of baby boomers retiring is going to create a glut in the housing market[2] as they attempt to unload the large homes supported by their working incomes for smaller, active retirement lifestyle accommodations. In fact, the average home size is expected to shrink back to mid–1990 levels by the year 2015, as more families lose ground in their battle for wage increases.[3] All of these news items indicate a trend toward higher-density, urban, mixed-use developments. Certainly cities are going to have to come up with creative solutions to address these challenges.

But the truth is, if you are still thinking of moving to the suburbs, you are not alone. Many fall prey to the false impression of cheaper land and lower rural tax evaluations. You should recognize, though, that as growth continues to move outward, so must infrastructure and services. And when enough people move outward, commercial development will follow to address the demand for support services and amenities. Before long, roadway construction will need to be upgraded or expanded to accommodate the volume of commuter vehicles; new schools, ball fields and hike-and-bike trails will be built; and strip shopping centers will line the highway. So even if property taxes are cheaper initially, that is probably not going to continue to be the case in the future. Some call this progress. Green building refers to it as sprawl, and the truth is that much of it is not sustainable.

Building new—or even sustaining existing—infrastructure for uncontrolled development is going to be difficult, if not impossible, except where planning addresses hubs of targeted growth corridors. We simply cannot afford to build and maintain an ever-expanding infrastructure to service every new "affordable" outlying development that springs up. Many cities around the country have partnered with their neighbors to create regional authorities to collaborate on long-term solutions that will enable them to stay ahead of these problems. By defining these target corridors, they are able to minimize traffic congestion caused by cross-area commutes from bedroom communities to work centers. Sustainable developments will have to be defined by planning and development boards that look at integrating employment centers, residential housing, parks and recreation both within the urban core and in suburban areas. To assure long-range sustainability, both affordability and diversity will have to be key components of these planned communities.

This is not to say that farmsteads are going by the wayside. On the contrary, there are also current trends toward community-supported local agriculture. We expect this form of agriculture will lead to increased activity for community-shared gardens, with many new opportunities in urban and suburban areas for small-scale organic cottage food industries. This will include continued efforts in the city structure of the future for developing vertical farms and embracing rooftop real estate for growing the food of those buildings' occupants.

Smart Growth

You can easily find new community resources (banks, churches, schools, retail, medical and other personal service providers) when you move to any new neighborhood, but finding them within walking distance, a feature offered by many of the new mixed-use "master planned" developments, offers even further savings on commuting expenses and time and other factors that impact your quality of life. In the best-case scenario, your location would be in a mixed-use development that includes all of the resources that your family will need: community services (fire, police and emergency) and recreation within walking or biking distance; alternative (car share) or mass transit (bus, commuter trains) that connects to other dedicated hubs. Since many cities are mandating Smart Growth[4] developments with these sustainability features, you should think about how well your investment will hold up in a location that does not.

Living in a master-planned development also means you get more amenities because the cost is shared by the community. This includes the cost of installing utility lines and roadways, open space, parks, recreational facilities including swimming pools, dedicated pedestrian and bicycle trails, libraries and community buildings and gardens. There is definitely a revival in interest in neighborhood support, with sharing of interests and responsibilities. Homeowner associations help assure properties are maintained and values are protected. To check the walkability of any site you are considering, visit walkscore.com. And, finally, when choosing your site, think about which one will contribute the most to your living enjoyment, building performance and ease of homeownership.

This chapter is dedicated to recognizing the opportunities and challenges associated with site selection. It is important to recognize that, although some sites may offer obvious advantages over others, sometimes if we think outside the box we can overcome challenges and still achieve our goals.

Analyzing Your Building Site

Green building uses strategies that embrace patterns in nature that create opportunities for cost savings. By incorporating passive strategies to utilize sun paths, capture prevailing breezes and water runoff, we can reduce dependency on mechanical systems. This allows us to design a comfortable home and only use those systems to augment nature as needed. To make the best of what nature has given us, we must analyze the microclimate of our location. How you build depends on where you are building; local climate dictates your best building practices. Climate is not just about region, though—it can be very site specific!

Whether you are selecting a lot in a subdivision, a condo in a high-rise, or a large plot of land, there are a few aspects of the building site that you should consider. We will discuss some basic building science in this book, so if you are not able to achieve a comfortable understanding of these concepts, then this is the point at which you should hire a building designer who can provide expert advice on which location might offer the best passive benefits, considering these three principles:

1. The sun rises in the east and sets in the west: the basis for passive solar strategies.
2. Hot air rises: the basis for natural ventilation strategies.
3. Water runs downhill: basis for water management strategies.

These three natural laws and how they exist on your site can create challenges or opportunities for building your green dream home. Unfortunately, land developers who plan lot orientations and plat lots cut out of large parcels of land do not normally put much thought into how those plot plans contribute to good home orientation. So, what to do if the building lot faces the wrong direction or has its long sides running on an undesirable axis? These circumstances are definitely where a good green architect proves their value. Designing for the site is never more critical than in these situations. If you already have a less than ideal building site, a good designer can find creative ways to take advantage of or overcome orientation issues.

The Sun Rises in the East and Sets in the West

Selecting a site that has good solar orientation is one of the best strategies to lower your home's operating costs. To determine the best way to orient your house based on this sun path, you need to

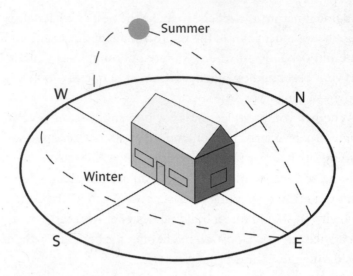

FIGURE 1.1. Determining the best orientation for the house on your site.

figure out where you are on the planet. In the northern hemisphere, the sun is *never* in the northern sky, and it is either high in the sky (in the summer) or low toward the southern horizon (in the winter).

We'll take a rectangular box design to illustrate the best design for passive solar benefits. To achieve the best orientation of your home on the site, you'll need to determine the sun path across the site, from east to west. Once you have charted the sun paths, you will need to determine if it is possible to design the home on the site with the long walls of the rectangle running east to west ("east-west axis").

It is important to recognize that the correct orientation of your house isn't limited to which side of it faces the street. Which walls are the long or short walls will depend on how you design the home for the site. This could mean the front is a long wall if your site is facing south, or if your lot faces west, it could mean that the side wall facing south is now the focal point of a courtyard. Or maybe the house faces north and the back patio has large overhangs that serve as a passive solar feature. But if the site is on the north slope of a hill, having good solar orientation for your home is going to be a concern. The point is to make the best use of your site by designing for it.

Hot Air Rises

Hot or warm air is less dense, causing it to rise above denser, cooler air. To take advantage of this phenomenon, look for a site that is in the path of natural wind currents. Take a few minutes to visit the National Resources Conservation Service website[5] to find data on the prevailing wind direction and average wind speeds for your area. Determine whether the topography of the site is going to promote airflow or prevent it. Is your site flat or on a hill? If it is on a hill, it's best to build on the side of the hill facing the direction that the wind comes from. So if your area's prevailing breeze is out of the south, you should be looking for either a flat building site unobstructed from the south, or the south slope of a hill. You don't want to build a house on the opposite slope of the hill, where stagnant air will trap heat or cold pockets, not allowing it to move on.

Water Runs Downhill

Next, look for natural drainage patterns when investigating potential building sites. If you take a close look at the topographical map of the building sites below, some of the lots have low spots in the middle, probably about where a house would sit. Having water coming downhill toward your home's foundation from all directions, even with the slightest slope, is not the preferred drainage plan and can lead to lots of costly problems in the future. If this is your situation, it will require designing alternative drainage routes and installing stormwater management controls.

In Chapter 8, we'll discuss the importance of managing stormwater onsite. For now you just need to make certain that you are selecting a building site where it will be possible to establish positive water drainage away from your foundation. Verify that the site does not have flooding or access issues, or is not in a valley shaped like a bowl, making it impossible to manage stormwater drainage, or at the least, very expensive.

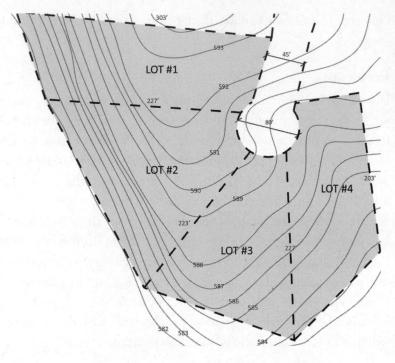

FIGURE 1.2. Site topography: Water flows downhill.

To take this concept a little further, though, it's important to note that water also runs down a roof and down the wall and basement assemblies too. So in designing the home, make sure you establish a drainage pattern that provides a path to get water off and away from your home in rain and wind-driven storm events. Thinking ahead at this early point in the process will pay off many times over later in money saved on basement and foundation water management and on maintenance to repair water damage to your home.

Your site management plan should include effective temporary (i.e., during construction) erosion controls and stormwater management. You don't want to lose your valuable topsoil that is exposed during the construction of the home. Silt fencing has long been recognized by building codes for meeting temporary erosion control standards, but other more natural materials work as well. Many green builders use straw bales to line the lot.

Other Considerations for Site Selection

Risk Assessment

At this point, it is important to do a risk assessment of your building site. Areas prone to flood, earthquakes, tornados or hurricanes, or even pests like termites can incur considerable structural damage over a building's lifetime. We like to think that most of the new buildings being constructed will stand at least 75–100 years, but sustainable construction teaches us to think in terms of hundreds of years. So, in analyzing certain risk possibilities, think in terms of what events are likely to occur over that time span. Hence the relevance of terms like "100-year flood," for example, when considering if the site is in a flood plain. It does not mean the event will happen one hundred years from the time that you build. Take a moment to think about the last time such an event did occur at the site. It might be that the odds are stacked against you: the event might be likely to occur sooner rather than later. If the area has seen increased development and impervious cover, you can expect more floods even with less rain in the future. Regardless, plan and design for such events, and specify materials and methods to mitigate any damages that they might cause on your site.

Many areas now allow development in flood plains, mandating stormwater management plans to mitigate damages and taxpayer-subsidized flood insurance for when those expected flood events occur. In essence, this means that we are committing the resources of the future to rebuild what we expect to be destroyed. And as this would be a reoccurring concern in these areas, it will also continue to be a losing proposition.

Some risks, like termites, are not assessed by historic frequency. Termites migrate in colonies, often having been introduced to an area on building materials (or even firewood) brought in from other parts of the country or world. The US Forestry Service map designates areas of the United States in terms of termite risk, from "none to slight," "slight to moderate," "moderate to heavy," and "very heavy." After determining which area you are in and what your risk tolerance is, it's best to plan accordingly during design and

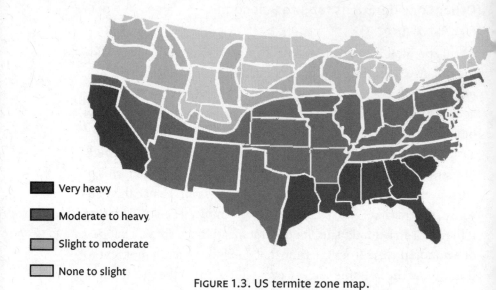

Very heavy

Moderate to heavy

Slight to moderate

None to slight

FIGURE 1.3. US termite zone map.

construction to minimize damages from this risk. Materials and methods to address with various risks are discussed in Chapter 4.

Site History

In choosing a building site, be aware of areas that have been subjected to pollution contamination. The Environmental Protection Agency's Brownfield[6] and Superfund[7] Programs require that identified disposal-contaminated areas are cleaned up before redevelopment can take place. Note the term "identified," as numerous sites have not been officially inspected and recorded, so be certain to look closely at the site and ask questions on its history. Only in recent history have we had regional legal landfill sites, so prior to that, and still in some areas, waste disposal has been by whatever means humans come up with.

Weather and Soil Stability

Construction methods and costs will vary considerably based on the stability of soils on your site. Planning for the changing stability of soil structure that can be caused by storm events can add complexity and cost to your project. Consider the site's accessibility in winter storms and during heavy stormwater events and the long-term costs associated with those issues.

California has areas prone to mud slides, the Mississippi River basin is prone to severe flooding that can sweep homes right off their foundations, frost heave results in tens of millions of dollars in damage every year in cold climates, and the Gulf Coast has drought-prone areas with expansive clay soils that face a continual threat of foundation problems. Working with professionals can help assure that your home is constructed with features to help minimize damages from the events likely to occur over its lifetime.

Protecting the Environment

Selecting a building site also requires us to analyze the impacts of developing that site on the environment. It is important to think about what the site will look like after we have completed our project. For example, imagine how the added impervious cover (concrete foundations, driveways, and other hardscapes) might increase stormwater runoff, causing erosion and limiting onsite infiltration. Other considerations affecting the environment include:

Keeping our Water Resources Safe: Anything that we do that impacts how water drains off the site might impact our neighbors and neighboring ecosystems. In many areas, we are now allowing development in the critical recharge areas of our surface water systems and groundwater aquifers. We do this even while knowing that the resulting pollution from stormwater runoff will carry lawn chemicals, fertilizers, pesticides and auto pollutants into the water our families will drink tomorrow. Just because these areas are now allowing development does not mean that we, as informed consumers, should support this effort. When you perform your due diligence investigating potential building sites, if you find issues like these, you should choose wisely.

Protecting Prime Farmland: Food shipped from around the world is generally grown on large production farms in soils that have been depleted of their natural nutritional value and that often use synthetic chemical fertilizers and pesticides. These foods must be harvested before they are ripe and often use chemical processes[8] to artificially ripen them in transit to our grocery store. The nutritional value of food has declined as a result of these practices. Additionally, this combination of industrialized food production and imported produce has contributed to the decline of small-scale farming as a profitable enterprise in the US.

It is unfortunate that much of our prime farming areas have been consumed by sprawl. This is due to the fact that many of our cities were located near the richest soil and most profitable farming areas in their region. Developers seeking cheap land to attract entry-level homebuyers have found that some of the least expensive land available is that farmland. At the same time, corporate commercial farming is driving small family farms bankrupt. Communities need to create incentives to conserve our suburban agricultural land to grow local, organic foods to keep us healthy. In the meanwhile, you can do your part by not buying a building site in one of these developments.

To make up for the farmland already lost to expanding development, we need to provide conservation areas of land in the urban core, from our own back yards to community gardens. To support this effort, we need to create organic waste management structures in our communities to provide natural, organic fertilizers for this purpose. These efforts are part of the core foundation of sustainable communities. We will talk more about how you can contribute to this effort in Chapter 13.

Where Did Nature Go? Talk to anyone in the town that you live in and ask them what that place was like just 25 years ago. If you can find someone older, ask about 50 years ago. You will no doubt hear that, of course, there were far fewer people and "I can remember when the outskirts of town was much closer in than it is now." You will hear stories of places that aren't there any more, places people used to go that were natural. Inevitably, you will hear people talk about land that's in the city now that used to be a large farm, where their father took them picking fresh produce in bushel baskets for canning as a child. How many of our current grocers shelves boast ingredients that are so local?

Many developers purchase land dense with tree cover only to completely clear it so that they do not have to deal with potential lot buyers who want to save this or that tree or have to work around a tree that's in the way. The wildlife that once lived there has been completely displaced. Even now, some of our neighborhoods are overrun with wild deer because their territories have been turned to suburbs. Then we even move into the greenbelts to put in golf courses for lifestyle communities, and push wildlife to the limit on space. And we consider them nuisances! Why is it that we humans think we own and control nature? Stewardship is a far better ethic in our opinion.

Those developers, and ultimately homebuyers, still do not recognize the long-term consequences of this approach. We have seen species go extinct from destruction of habitat, while other populations explode from destruction of predators, both at the hand of man. Now we find ourselves forced to designate wildlife sanctuaries to protect birds and butterflies from loss of their migratory habitats to development.

We need to remember that we are part of a great food chain that starts with the smallest creatures and works its way up to us. We, and the meat that we eat, both eat plants. Plants depend on the worm that feeds the bird that fertilizes the soil that grows the plant.

We are all interdependent for our survival. This reminds us of a John Muir quote, "When we try to pick out anything by itself, we find it hitched to everything else in the Universe." It is also important for our own health that we still have places, albeit nature preserves or just little backyard sanctuaries or pocket parks, where we can still commune with nature, smell fresh air, see butterflies and hear birds sing. This topic will be discussed further in Chapter 8.

Payback on This Investment

To get a true picture of site costs and site value, we need to look at the larger picture. We use fossil fuel resources every day for commuting from our homes to our work or other activities. When we build in remote areas, we dedicate resources to construct and maintain (forever!) roadways, utility services and other services. Then we use even more fossil fuel resources to travel to the same activities because the commute is now much longer. This is one of the reasons that green building practices promote high-density, mixed-use urban infill developments.

Sprawl forces us to take on bonded indebtedness to pay for new roads, schools, water and sewage treatment facilities and new power transmission lines. This new bonded indebtedness will take years, if not generations, to pay off, with no end in sight for our continued thirst to add more! Our existing urban core infrastructure was built and paid for long ago at a much lower initial cost. Much of that infrastructure is now at least 25 years old (and some of it 75 years old) and in need of maintenance and repairs, but since our tax revenues are committed to pay off these newer bonds, funds are not available to do this. Many cities now face serious groundwater contamination from old leaky sewage and water delivery pipes they no longer have funds to maintain or repair. By increasing density in existing urban areas rather than creating more sprawl, we can take dedicate new bond funds to maintaining and updating existing facilities, addressing the additional service load capacity and reducing the waste.

The same is true for utility and emergency service providers. With high-density urban development, service providers can focus investments to increase the capacities within their existing territory, and servicing higher densities means that they are able to spread that cost over a larger customer base. But as developments get more spread out, providing services to the same population requires more investment in delivery systems and personnel, passing those added costs on to all of their subscribers. Sometimes developmental fees are charged directly by utilities or municipalities to these rural developments for the costs of adding additional infrastructure. But ultimately, the costs of adding more support and emergency services associated with sprawl, and conversely the cost savings associated within increased urban density, are reflected in the tax base of the community. Property taxes, sales taxes, or whatever form your local jurisdiction chooses to pass on these costs as well as portions of the federal income taxes paid by working taxpayers nationwide, all contribute to fund federal financial road and infrastructure assistance programs. However they are paid for, these costs will affect the bottom line of your total home budget. Not only now, but for as long as you own your home, even after the mortgage is paid off.

Let's look at some of the potential pitfalls for the long-range return on investment (ROI) potential for your affordable home in the country versus the urban infill bungalow:

1. Gasoline prices are expected to soar to $7–$8/gallon in the next couple of years. Along with wear and tear on your automobile, what does that do to your household budget? What do you think those costs will look like in 10–20 years? How about 40–50 years? Will those homes be abandoned then because the cost of commuting will no longer be feasible?

2. Populations are growing, more and more people are moving to the cities, often into the ever-expanding sprawl developments. Suburban sprawl creates long-distance commutes to work centers, shopping and recreational facilities, and is recognized as the main culprit for excess automotive travel and traffic congestion. How many hours a day will you be willing to spend commuting? What is that time worth, and what other activities are you giving up because of being robbed of that time?

3. Think about the impact on air quality related to pollution from urban traffic automotive exhaust. Stormwater runoff carries automobile oil and gas waste from roadways into surface and groundwater supplies, again affecting drinking water quality. And we all share the impacts of air and water pollution on our health and healthcare costs. Reducing the frequency and duration of automotive travel is key to reversing this trend.

4. Public service budgets are stretched to limits, the baby boomers are reaching retirement and will further strain Social Security and Medicare resources, meaning taxes are going up. Some municipalities and other government entities are already cutting spending on non-essentials, like non-critical maintenance. Where's the money going to come from when your roads and utilities need maintenance or repairs?

On the other hand, if we choose an urban infill location for our home, in a high-density, mixed-use community with dedicated pedestrian and bicycle transit alternatives that connect us to retail, medical and other personal service providers, schools and recreation, with mass transit options for when we need to commute, we minimize our impact on the community and the impact that the concerns listed above have on our quality of life. More and more cities are investigating building mass transit alternatives for commuters, since fossil fuel supplies are diminishing quickly, traffic problems are getting worse as urban populations increase, and these same cities are under scrutiny as being non-compliant with air quality standards set by the Environmental Protection Agency (EPA). While the cost of building and maintaining these transit systems would create additional tax burden on the entire community or region, it would be far less than the cost associated with uncontrolled sprawl and unbridled sources of air pollution.

When we choose sprawl development, taxes might initially be less expensive than for property in an urban locale, but your additional commute time and expenses alone should offset those savings:

Do the math for a working couple:

Taxes on urban infill home valued at $380,000	$8,000/year
Drive 6 miles round trip to work @ $0.50/mile (2 cars, gas, maintenance)	$1,500/year
Combined urban expense	**$9,500/year**
Taxes on home in rural subdivision valued at $250,000	$2,400/year
Drive 30 miles round trip to work @ $0.50mile (2 cars, gas, maintenance)	$7,500/year
Combined rural expense	**$9,900/year**

Taxation

Let's take a closer look at the cheaper property taxes. Suburban or rural bedroom communities still provide much of the workforce for their urban "mother" city. As this type of development grows, taxes must be increased to support infrastructure (roads, utilities, emergency services and especially educational facilities) that must be provided. Eventually those bedroom community tax rates catch up to their urban counterparts. This is because urban dwellers typically pay a smaller proportionate share of costs due to their higher density.

All Things Considered

So, thinking about where you are going to live should include consideration of its future affordability. Do not wait until gas prices are $6.00–$7.00 per gallon, or there are not enough hours in the day to fight traffic to shuttle your active family around in gridlocked traffic, and you see your investment value deteriorating due to these and other changing market conditions. You will be glad to have done your homework and made a better informed decision for your home's location.

Hmm, still can't decide? Before we go any further in choosing your location, we recommend that you take a few minutes to reflect on your life in terms of where the daily activities of your life occur geographically:

1. Look at a map of your city and mark an **X** at the general locations where you and your spouse go to work. If you and/or your spouse works from home or is not employed, do not mark your home location. Only mark locations where you must commute to work. If you are an outside salesman and travel a territory, mark the outline of your territory on the map.
2. Mark the location of each of your children's schools. Again, if any of your children are home-schooled, do not mark the location of your home.
3. Mark the location of other fixed-place activities that any member of your family is committed to. Only count places that, due to some contractual obligation, cannot be easily changed. Do not include the grocery, park, fitness facility, or shopping mall.
4. Mark the proximity of any close family relatives that provide vital contributions to your families' schedule.

Now it's time to determine the best location for you to live. To do this, ask yourself these questions:

1. Is there an acceptable location halfway between you and your spouse's employment?
2. Are there schools close to this better home location that would offer the same quality of education as those you marked in 2?
3. Is this better home location more convenient to the extracurricular activities you noted in 3?
4. Would moving to that better location negatively impact the family support you recognized in 4?
5. Would the move to the better location have an overall positive effect on your family's time and schedule?
6. If it is a more expensive neighborhood, would the cost savings of less commuting (gasoline, wear and tear and time) be enough to offset it?

A couple of colleagues did this analysis, one with a 6-mile round-trip commute and the other with a 24-mile round-trip commute. At $3.50/gallon for gasoline alone, over a 5-day work week, 50 weeks/year, the latter pays $787 more in fuel at 20 miles per gallon. Based on that alone, the mortgage payment on the closer home could be $65 more per month and the owner would have an extra hour every day not fighting traffic. Based on a reasonable wear-and-tear cost of $0.50/mile on the automobile, the second colleague spends $2,250 more every year for their work commute alone, or $187.50/month that could go towards a higher mortgage payment and more free time. That adds over $253/month to a higher mortgage payment for an urban home.

To sum it up, here's a breakdown of the additional costs for rural living versus urban living:

- Cost of building and maintaining roadways for commuting/connection from suburban to urban area
- Cost of installing and maintaining utility lines to rural development (phone, cable, water, electric, gas)
- Cost of gasoline for commuting and accelerated wear and tear on automobile from faster accumulation of miles spent commuting
- Value of your time spent commuting, mental and physical strain driving X hours per day

- Cost to build, staff and maintain fire, police, trash and emergency facilities to service rural areas
- Cost to build support services (convenience stores, gas stations, schools, childcare, churches, etc.) to service rural development
- Cost to build amenities (recreation, parks and green space, etc.) to support rural development
- Cost to install and maintain mass transit to service rural development (bus, commuter rail)
- Pollution related to high volume of commuter traffic from rural areas to urban areas, cost of cleanup, higher insurance premiums and healthcare costs
- Loss of valuable agricultural land, wildlife habitat, and nature areas to rural development

We hope that you have read this chapter before you begin your search for a suitable building site. If not, we suggest that unless you are retired, work from home or won't be commuting daily to work from your site over 15 miles one way, you should consider postponing your build until you are in a position to do that. Or, if you could not pass up the current buyer's market and low interest rates, you could still build now and rent it out to others who would be home-based until you yourself are able to stop the daily commute.

CHAPTER 2

Size Matters

Anyone can tell you that building a large house will cost more than a small one. Every additional foot of space in every room takes more labor and materials. More space adds more foundation and roof, and more stone, brick or siding. Larger homes take more structural materials to build, more interior finish products (cabinets, flooring, drywall, paint, windows, doors), larger systems for heating and cooling and more furnishings. Large lots require more landscaping. Whatever your budget, the larger the space the larger the percentage of funds will go to base construction cost, leaving less for high-performance features and amenities.

A common misconception is that larger homes are more economical to buy, with an analogy made to volume purchasing, where the price per square foot is lower than for similarly built smaller homes. We often hear salespeople say that they get better pricing from their subcontractors for larger homes, because "they are already on the job" so adding more square footage does not significantly increase their bids. The truth behind square-foot pricing is that some of the larger ticket item costs are basically the same regardless of home size, like the kitchen, the water heater and a couple of bathrooms. Since these costs are averaged out over less square footage in a smaller home, they have a greater impact on the average square-foot price.

So let's take a minute to think about this whole idea of valuing homes by their square footage cost. Builders have figured out that by building larger homes they spread these fixed costs over more square footage and get that per square foot cost down to some number that they think will impress potential buyers. Two-story homes are also less expensive per square foot, because they need half the foundation and roof size of the same home built as a one story.

However, the flip side of that value equation is that most property taxing authorities use home size as one of their main calculations to determine appraised values. Typically, the values of comparable properties (called "comps" in the industry) of similar style, construction and age are used to determine this square foot value. Of course, the larger your home, the more you should expect to pay in property taxes.

Pricing homes by the square foot for comparison purposes just does not make sense. It is no different than pricing our automobiles by the cubic foot of cargo space, when cars can differ widely in quality, amenities and finish. In fact, even when we buy certain appliances (refrigerators, water heaters) by their capacity, we don't calculate their cost per foot, we just compare their total cost, features and benefits.

Just as we are now acutely aware of what kind of gas mileage our cars should be getting, we should also plan for how efficiently our home will use our energy and water resources over its lifetime, as those resources are also being depleted and their costs rising. We certainly hate wasting these resources, but many are wasted due to poor construction, poor system design and poor resource management. We should be building smaller homes with highly efficient systems designed not to waste resources.

So be wary of getting caught up in square foot pricing. Our goal should be to get the most beneficial features possible in the smallest area that meets our needs. Look at real, usable space, and amenities, amenities, amenities. Use the money that you save on cutting down the size of the structure to spend on better systems, equipment and interior finishes. We will provide guidance on those choices in future chapters. Just remember to keep your total budget in mind knowing that saving money here will keep that money available to you to have later when you make those selections. When it comes to size, think in terms of inches (instead of feet) and maximize the space with classic features that will last the test of time.

If you've never seen a copy of Sarah Susanka's *Not So Big House*, we highly recommend that you invest in one. Full of insights and incredible photo ideas, it helps you realize the true value of space and creating a design that represents how you really use it. For even more information and references, visit her website at notsobig house.com. You should really start here *before* you think about a floor plan.

Assessing Your Needs

Take time to anticipate your current and future needs in designing the home. How large a house do you need? You should start this thought process by sitting down and making a list of your family's current and anticipated future needs in housing.

Changing Needs Over Time

Are your kids still at home, and, if so, how many more years would you expect them to stay? What are you going to do with that space after they are gone? Could you design that space to serve some other purpose, to continue to be usable, valuable square footage, or maybe even design it to eventually become a stand-alone apartment? As our family dynamics change over time, our needs for how the home functions will also change over time. It is important that we design a home that will serve our needs over the long term.

In Chapter 1, we talked about homes built in the US currently having an average life expectancy of at least 75 to 100 years. If we think about who might occupy the home over that time period, we realize that many of the decisions we make when constructing it should be expanded to include considerations beyond our current wants and needs. This not only makes better use of resources over the long term, it also protects the resale value of your home.

Single adults will soon be raising a family. Their tiny babies become crawling toddlers, rambunctious kids and then large teenagers with herds of friends and, eventually adults themselves, moving out. The parents, now empty nesters aging in place, find that the abandoned second floor of their two-story home continues to impact their utilities and property taxes and, therefore, their retirement savings. This may cause them to have to move away from their beloved neighborhood in order to downsize.

What about later in life, when your parents are older and might need to come live with you? Do you have a space that could function for them? Or as you yourself age, is there a live-in apartment for a caregiver? Maybe that space might serve to supplement your income in the interim to help pay the property taxes, as this may provide the means for you to afford to stay in your home.

How well will your home adapt to these changes? Can you design a home that will have a flexible enough design to conform to these changing needs over time? Many young couples buy homes in anticipation of a growing family, and then as empty nesters often do move to a different neighborhood in order to downsize. This is because we often fail to think of the other possible scenarios and thus fail to act accordingly in our planning. As the old proverb says, "We never plan to fail, but we often fail to plan."

Accessibility

If you have ever spent any time on crutches or in a wheelchair, you have gained an awareness of how accessible areas of your home were (or weren't) during that period. Now is the time to think about some simple design specifications that would allow you to live more comfortably in your home if your mobility were ever temporarily or permanently compromised. Are the passageways wide enough to be navigable? Can you get into a downstairs bathroom with room to turn around? Can you get in and out of the shower or tub without losing your balance?

Think about having your home designed to meet the Americans with Disabilities Act.[1] This act does not apply to private residential structures (except some that may receive certain federal funding), but can be useful as a guideline for designing and installing features to address these issues. Again, think about who might occupy the home over its hundred-year life. Certainly someone during that period will benefit from your foresight, possibly increasing your resale market value

Many accessibility features are easy to incorporate into new construction and usually do not add any additional cost to the budget on a new home. Yet they can be very expensive and difficult to retrofit if added afterwards. These include blocking inside the walls in showers and tubs that enable the easy installation of grab bars in the event they are needed; electrical outlets installed no lower than 18" and light switches no higher than 48" above the finished floor; a no-step door threshold that provides easy entrance for a wheelchair; and wide hallways and doorways to allow navigation through the main living areas and first-floor bath. Even if you never have a temporary injury, these features become more valuable as we age.

Flex Space

Flex space can mean a lot of things. It can define a space in your home that can serve more than one use at the same time or can be transformed to different uses over time. For example, consider a home that is occupied 99 percent of the time by empty nesters. They look forward to having their family home for holiday feasts, but having a formal dining room just does not make sense for a number of reasons. With limited time to visit while preparing a meal or cleaning up, having an eat-in kitchen increases their time spent together. This arrangement also suffices for the occasional friends-over-for-dinner affair, but does not accommodate a large family function. What if we designed the home with a large cased opening between the dining space and the living room? We could buy a gate leg table and some folding chairs that can be set up to add an additional six seats to our existing table. This functions wonderfully for the whole family to enjoy holiday meals together but doesn't create a wasted space the rest of the year.

Think about innovative ways to use every nook and cranny of space within the exterior walls. Do you have a space that can serve as a home office if needed, or maybe a hobby room? This could be an open area adjacent to the kitchen that could also be used as a formal dining room or guest room. Again, think about how future homeowners might use the space and its resale value.

Flex space can be in the form of built-in bookcases lining stairwells, or an under-stair day bed that functions as reading nook or guest bed. Or it could be a Murphy bed in the study to offer guests more privacy, a folding table in the laundry room that converts to a craft table or has a sewing cabinet built in with cabinets above that store office, sewing and cleaning supplies.

There are lots of great examples of creative space design. We recently visited a new home where the owner had designed a pass-through closet from the master bath to the laundry room, with direct access to the dirty clothes hamper and shelving for towels. Many walk-out basements are pre-plumbed for later conversion to apartment-style housing for supplemental income or an in-law suite. Or a garage apartment could also be a home office or temporary housing for the teenager off to college transition.

Operational and Maintenance Costs

When analyzing home costs, it is important to recognize the overall cost of ownership over time. Think about home size in terms of operational and maintenance costs. Certainly it will cost more to heat and cool a larger space. As energy costs continue to rise, larger homes especially will face even higher utility costs in the future. The same is true for homes that are not plumbed for efficient use of water.

So think about how large a home (and yard) you want to maintain. All homes require maintenance, whether periodically repainting the interior rooms or trim, or replacing a roof. The larger the home, the more it will cost to maintain. The larger the yard, the more landscape maintenance, mowing, weeding, watering and leaf raking.

Think about the home's durability: how long the materials used to construct it are expected to last and how often and what kind of maintenance, upkeep, repairs and replacement cycles are expected of each component. Are you planning a home that you will be able to maintain in the future? Certainly, as we get older and become less able physically and economically to do these tasks, having less maintenance and fewer repairs is more important to us. And as resources continue to become scarcer, materials for repairs and replacement will be more expensive as well.

Furnishing

Larger homes require more furniture, whether it goes in additional rooms or larger spaces. But more furnishings also include more window treatments, more lighting, more floor coverings, more decorative accessories, even more kitchen gadget appliances. When we have more storage space, it tends to make us want to buy more stuff: more clothing, more holiday decorations and more toys. Many of these products contain toxic materials that impact indoor air quality and our health (more on this in Chapter 7). And all of these material possessions are further depleting our natural resources and contributing to environmental degradation. Over time they wear out and must be replaced, as those large spaces continue to dare us to fill them, creating a whole new cycle of impacts.

Cleaning

Think about how large a home you want to clean. It takes time, energy and water to clean, and most off-the-shelf cleaning products contain toxic chemicals that affect your health. Even if you use nontoxic natural cleaners, having a larger home certainly means that you have more to keep clean and it will cost more time and money to do so.

Paybacks on Right-Sizing Your Home

A huge dining room for large family holiday meals may not be the best solution when you realize the expense of building, heating and cooling that additional area, as well as keeping it clean year round. Then there's the investment in that large dining table and twelve chairs that will only be used once or twice a year. The same is true for the large family room you thought might be the best solution for spending time with your children when you realize how much furniture it takes to fill it up and how that might result in separate seating areas that create spaces apart from each other.

Build a home to meet your needs. We don't know of anyone that *needs* a McMansion. Building a smaller home can save you money in both initial construction costs and in operational and maintenance costs over its lifetime. The savings from reducing square footage should be used to build a more durable, well-designed and better insulated home with high-efficiency systems and healthy home finishes. The paybacks on this combined strategy are beyond description—it just doesn't get any better! Some are determined by net energy and water consumption costs, while the value of your family's health and that of our planet cannot be quantified. Green building is about supporting sustainability at every level, including conserving resources and keeping our environment healthy, but in the final analysis, it's really about how these things affect your quality of life.

CHAPTER 3

Design, Design, Design: Everything You Need to Know

When you think about design, there are three aspects of a green home that you should remember. The first two aspects of design were introduced in Chapter 1:

1. Design in terms of site: views, topography, and natural resource inventory.
2. Design in terms of passive systems: passive solar, natural ventilation, drainage.

In this chapter, we will look at those in more detail and introduce the third strategy:

3. Design in terms of materials efficiency.

We are not saying that design is everything, but it is your single best opportunity for natural resource efficiency and saving money. It's as easy as 1, 2, 3. Really!

1. Designing for Your Site

In Chapter 1, we looked at site selection in terms of natural amenities, topography and location. Our site might have other characteristics that influence our design. Desirable views (or objectionable ones) certainly affect site-specific design. Natural site features can be used to improve building performance or, at least, define how we utilize the site to accomplish that with installed components. When we consider the design for our home, we should assess the natural inventory of the site and take advantage of any unique opportunities that it presents.

Site Management

This is really important to think about *before* you start construction on your land. If you can limit the amount of site that you disturb during construction, you will save money not only on site work, but also on restoration work and installing your landscape. If we truly want to keep housing affordable in the long term, we must consider eliminating all costs that do not add benefit, especially those that increase construction costs and long-term maintenance, increasing our total cost of ownership. There is no need to feel like you have to do something with every inch of your lot. Sometimes less is more.

So this is the time to set your site construction parameters and create a plan for protecting the land outside of the area that you will allow to be disturbed. Consider fencing off that protected area with orange plastic construction fencing, or maybe using straw bales not only to define that space but also to reduce erosion of native soils that are disturbed during construction. This effort is especially important for protecting any existing trees during construction.

Foundation and Roof Design

The type of foundation system you select should primarily depend on what is best suited or required for your building site and climate. For sites with poor soil stability, you should consider deep piers or beams that will connect the structure to a point of ground stability like bedrock or below the frost line. Even if your site does not present those concerns, it's best to have your home's foundation designed for your specific site by a qualified engineer. Otherwise, you may overspend on more foundation structure than you need, or end up paying more in the long run for repairs because the foundation was not designed to handle the stress specific to your site.

If you are on solid ground, you may have the option of choosing between a slab foundation, a basement or some type of pier and beam support structure. Pier and beam systems give you long-term access to any systems that run below the floor of the home, including plumbing lines, sewage lines, gas lines, etc. Over the life of your home, you may need to repair or replace one of these service lines, and this is much easier if you can access it than having to jackhammer out a solid concrete slab foundation. Pier and beam and basement systems also give you the benefit of easily relocating any of those system lines in the event you decide you must reconfigure room arrangement or use.

If you intend to build a basement, it is imperative that issues like bulk-water intrusion, rising groundwater, radon and condensation be addressed in the plans *before* construction begins. A basement can be a boon or it can become your worst nightmare. The latter outcome is usually due to water, so how this is planned for and handled is the key factor that separates one outcome from the other. We will address the building science of how to build a dry, healthy basement in Chapter 4. For now, let's just say that it's no longer true that a basement must be a place with musty odors and occasional standing water.

Sloped sites might require foundation walls that take large quantities of materials to construct. You may need to think about whether or not skirting is really necessary on your project. If it is, to reduce the amount of materials needed, consider stepping the foundation down grade changes or supporting a level floor on a permanent, durable stilt structure. Alternative building components may offer a better value, such as insulated concrete forms (ICFs) or structural insulated panels (SIPs). These systems work very well if you are installing a sealed (unventilated) crawlspace or basement (discussed later in this chapter).

Your Home's Umbrella: Roofs are more than just architectural features. The main consideration for roof design is more than just the longevity of the roof itself; it also serves to shade and shield the

Moisture Issues in Walls vs. Roof Overhangs

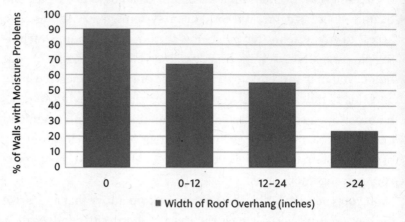

FIGURE 3.1. Moisture issues in walls vs. roof overhangs. Credit: Adapted from HUD/NAHB, Office of Policy Research and Development.

wall components from sun, rain and other natural weather events, like snow or hail. This means the roof overhangs should provide protection for the home's exterior walls, doors, and windows. You should assess the area of your site where you intend to build for sun exposure and any protection offered from other nearby features.

This is a big deal when it comes to your home's durability. Left unprotected over time, door and window frames can leak and walls can suffer water damage and mold. Keeping that water out to begin with is one of the best functions that your roof can provide. Research has shown that the deeper the roof overhangs, the lower the chances of water intrusion on the walls of the home (see Figure 3.1 above). The hip roof design (roof sloped to all four sides) offers the best overall protection.

 Solar Ready: If your lot is not too heavily tree-shaded, designing the roof with the ridge running east to west provides plenty of space for future installation of solar panel components. Whether your plans are to generate onsite energy through solar photovoltaic (PV) systems or provide hot water with a solar thermal system, the roof solar access is critical. To mount the more common type of solar arrays that operate off a common inverter on the roof, you'll need approximately 100 square feet of roof per kilowatt (1 kilowatt =1,000 watts) of array. (Micro-inverter panels do not require series installation, so although they are more expensive than series-inverter arrays, their placement is not affected as much by roof penetrations or continuous roof space.)

Later in this chapter, we will discuss energy modeling, which is useful in determining the size of solar array you'll need to manage your electrical loads. For now, just remember that you will need that amount of roof area with the proper orientation and roof pitch in order to maximize your solar PV production. This area needs to have a clear southern exposure (no shading) with minimal or no penetrations (roof/plumbing vents). Roof vents are usually required above any combustion appliances (furnace, gas water heater, cook top), as well as over bathrooms, kitchens and laundry rooms (plumbing and ventilation pipes). With direction and planning, contractors can find alternate locations for roof vents, on adjacent roof areas or even exterior walls. Make sure that you communicate your intent to keep roof space dedicated for future solar access. For more details on Solar Ready planning, visit nrel.gov.

Even if you do not plan on installing any solar devices in your construction project immediately, think about the life of the home. Most homes built today in America will still be standing long after fossil fuel energy sources are completely depleted. You may realize in as little as five, ten, or fifteen years, that having the roof design such that it will allow you to add onsite active systems is a great hedge against rising costs. And certainly over the course of a hundred years, other owners and occupants may value that the home's roof design allows that option, which will help in resale value.

Avoid chopped up roof designs. The most basic, cost-effective roof design is a straight ridge running east to west (a gable roof). Although a complicated roof design might add architectural interest, every turn and valley increases the potential for water leaks, which adds cost in flashings and other materials to keep water out and to ventilate the assembly. These costs are incurred not only during the initial construction, but every time the roof has to be replaced. Also, the more complex the roof design, the greater the number of areas for water to eventually penetrate, thus increasing the frequency of costly roof repairs and/or replacement.

Vented versus Sealed Attics and Crawlspaces: Building science research has made many recent advances regarding ventilation strategies for attics[1] and crawlspaces.[2] These studies indicate that when the HVAC or ducts are located in the attic, unventilated assemblies outperform their ventilated counterparts in all climates, which saves you money over the life of the home. And, yes, these types of assemblies are approved by the international building codes. We now know that in hot, humid or mixed-humid climates, ventilated assemblies can cause a number of issues in the home. In a humid climate, any vent openings in your attic and/or crawlspace can provide a path for outdoor humidity to enter the cool crawlspace and cause condensation; in the case of your attic, the extreme heat forces your cooling system and ductwork to work in a very hostile environment. If your attic and/or crawlspace are ventilated with this humid summer air, it allows that humidity much greater access to your home via all of the many penetrations. Vented crawlspaces should only be built where they can be ventilated with dry air for most of the year.

The alternative to ventilated attics is to seal up these spaces, just as you seal up other parts of your building assembly, and install insulation at the exterior of that sealed space. This means insulating with spray foam at the roof deck and rafters, just as you would at all other exterior walls of the structure.

The unvented attic is a clear winner if you intend to have any mechanical equipment or ductwork in the attic, as is often done in warm and mixed-climate zones. By sealing up the attic, applying spray foam insulation to the underside of the roof deck and rafters, you bring that system within the air boundary and thermal envelope. In other words, the system components are inside the insulated air-sealed space so they are protected from heat in the summer and cold in the winter. In a traditional vented attic, temperatures can exceed 130 degrees in the summer, so placing our air conditioner equipment and ductwork there is like putting them in an oven and still expecting them to keep our home cool. The same is true for our furnace (and its ductwork) if located in a vented attic, having to overcome winter's cold while trying to heat our home.

Many times locating the equipment and ductwork within the insulated space means that we can downsize the size of the system. That saves you money initially on the installation of the system. You will also save money on utility costs over the life of the home since a smaller system will typically use less energy to operate. This operational savings will be even greater because of the decreased heat or cold loads on the equipment and ductwork that results from locating them inside our insulated house, further increasing its efficiency.

In cold and mixed climates, the foam insulation effectively prevents ice dams by keeping the roof deck from becoming warm enough to melt the underside of the roof snow pack. We now know that the melting of the underside of a snow pack caused by rising heat from the house is what begins the process of ice dam formation. This meltwater runs down to the cold exposed eave and refreezes to start the process of ice dam formation. A sealed attic assembly eliminates the need for heaters at the eaves or working on dangerous ladders to control the ice buildup at the overhangs, since the snow pack never melts and the ice dam never gets a chance to start. Installing foam insulation also prevents hoar frost because there are no surfaces in the attic cold enough to be a condensing surface, much less have frost form on them. Building science gives us the power to address home problems at their source, rather than trying to control the symptoms after they have begun.

The other option is to go old school and ventilate your attic. The new twist is that we now know that if we do this we must create an airtight seal with foam insulation or other rigid materials (wood or drywall) at the ceiling of your home. If you choose this option, you must take special precautions to ensure that the entire ceiling plane with all of the framing irregularities like utility chases, dropped ceilings, furrdowns, can lights and trey ceilings is fully air-sealed. The key is to cover and air-seal all these gaps and irregularities. If you employ this option, you can't efficiently use the attic for your ducts or mechanical equipment, so you should consider placing your ducts in dropped ceilings below the attic insulation and air barrier. This will allow you to still reap the benefits of your mechanicals within the thermal and air barrier of your home.

The best venting system to use includes both continuous ridge vents that run along the top of the highest ridge points of the roof and continuous soffit vents that are placed in the eave underside of the roof overhangs or otherwise at the base of the roof assembly. This combination uses the free physics of a thermal chimney effect to draw the hot air up and out of the attic and replace it with the cooler air from the shade of the eaves. Vented crawlspaces should only be built where they can be ventilated with dry air most of the year.

Roof Truss Design: If you are designing a home with a vented roof assembly, it is critical that the design include details that will allow the home to be constructed for the best thermal performance. Remember, all homes should provide the best possible shelter from extreme weather conditions, but a green home should strive to achieve this by using good applied building science, better building methods and sound, passive design strategies. These efforts result in a durable, high-performance home that will continue to provide operational cost savings throughout its lifetime.

Building better requires us to look at how effectively traditional building practices have performed and whether we can improve upon those practices. Roof truss design is a good example of this. In traditional wood truss design, the roof rafters sit right on the ceiling joist, so there is not room to take full-depth insulation all the way out to the edge. This means that we end up with a fairly large gap between the end of the ceiling insulation and the top of the wall insulation, along the soffit line. In a traditional wood-framed home,

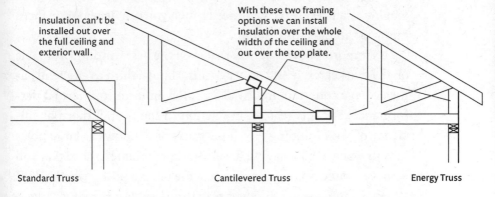

Insulation can't be installed out over the full ceiling and exterior wall.

With these two framing options we can install insulation over the whole width of the ceiling and out over the top plate.

Standard Truss   Cantilevered Truss   Energy Truss

FIGURE 3.2. Truss designs.

this is one of the most commonly missed areas of insulation, and since hot air rises, it provides a good escape route for heat to move outward from the conditioned space to the outdoors in the winter (see "Thermal Bridging" in Chapter 4). In cold climates, the ceiling gets cold in this area, causing water condensation that leads to mold growth.

A better way to design trusses is with energy trusses or raised heel trusses. Energy trusses or trusses designed to cantilever out over the top plate provide an opportunity to get the insulation out all the way over the top plate of the exterior wall assembly. If your goal is net zero energy, the improved design and thermal performance work together to reduce loads on mechanical systems, i.e., building energy use. The International Energy Conservation Code also recognizes the inherent energy performance benefits of energy and cantilevered truss designs by allowing you to reduce the overall attic insulation level if they are used.[3]

Fenestration (Windows)

Glazing: Where you put windows, commonly called the "fenestration" or "glazing area" by building designers, should have much more to do with their function than with the aesthetics of the design. Unfortunately, in traditional home design, this is seldom the case. Try to think of windows in terms of their usefulness when placing them in your design. Once you have determined how the windows can best serve your needs, then you can incorporate them into your architectural detail.

It's important to recognize that even the best windows available on the market today still only perform about a third as efficiently as the average code-required wall assembly in terms of thermal performance. For this reason, in relation to the overall thermal efficiency of your building envelope, think of windows as large holes in your walls. While having fewer windows is better, those that you do have should serve one or more of the purposes described below. When planning your windows, plan to accomplish as many of these benefits as you can with each window.

Since the sun's path is east-west, you will want to design so that most of the windows are on the sides of the home that support or reduce passive heating or cooling, as appropriate for your climate. If you live in a hot climate, you will want to reduce the amount of window glazing area on the east and west sides of the house. This will help you save money on your summertime cooling costs. If you are trying to maximize heat gain in a cold climate, then you would want most of the windows on these sides, as well as the critical south side.

Also remember that windows are available in a variety of shapes and sizes. The only reason to install windows in any location that you expect to keep covered all of the time with privacy treatments is to provide an emergency exit in case of fire or other interior threat. Don't get caught up in thinking that all your windows must be three feet by six feet or four feet by five feet (in construction jargon, 3060 or 4050). Instead, think about how you can use the light. The same amount of glass can be installed in one large window or several smaller ones. Placing a few two-foot-by-two-foot windows at standing height eye level would give you nice views, while also providing daylighting, and yet would not interfere with furniture placement or add privacy concerns. Adding a lot of windows where they serve no purpose is just wasting your construction budget and will also increase your utility, maintenance and repair costs over the life of the home.

Views: Windows can perform the important function of connecting indoor space to outdoor space and capturing any views that will help you stay in touch with the natural environment and events outside. This is important not only for your health but also security, by letting you see what is going on outside the home. However, that works both ways—windows added as architectural features may serve no other purpose than providing unwanted intrusions into your privacy.

2. Designing for Passive Systems
Passive Ventilation

Repeat again: ventilation. We tend to believe that all windows should be operable. Regardless of where you live, there are times of the day and times of the year when the outside temperature and humidity are comfortable. Fresh air can provide numerous health benefits, and the sounds and smells allow us to be a little closer to nature, even as we stay inside our sheltered home.

In Chapter 1, we talked about the importance of determining the direction of prevailing breezes and selecting a building site that could take advantage of passive ventilation strategies. These strategies should be a part of every good green home design. But for those breezes to really provide cooling for the home, the design must incorporate elements to promote air circulation through the building.

Windows, depending on location, can be used for cross ventilation, circulating natural breezes throughout the home, and stack ventilation, exhausting heat out through stairwells, monitors, cupolas or upstairs rooms. However, prevailing winds can be impeded by a number of factors, including adjacent structures and topography that either block the airflow or change its direction at window level. But if we can get above the obstruction, we can use our roof design as a scoop to channel the wind back down into the building. Wing walls can also accomplish this at ground level, if we can capture the breeze coming around the neighboring impediment and direct it back toward our other architectural features that

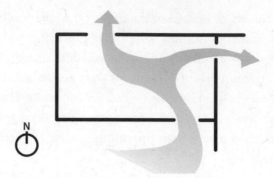

FIGURE 3.3. Cross ventilation aided by wing wall.

are designed to capture it. Investing in these kinds of architectural details will pay off over the long run by reducing your mechanical cooling costs.

We can also use natural ventilation strategies instead of running our air conditioners, thereby lowering our utility costs. Open windows can reduce our need to use mechanical ventilation appliances, like bath and kitchen exhaust vents. Opening the windows can also improve indoor air quality by exhausting stale, contaminated air, allowing us to temporarily shut off our mechanical fresh air systems. Opening the laundry room windows can provide a source for make-up air for the clothes dryer, preventing negative pressure that might bring in air from unwanted sources.

It is important to note, however, that operable windows should not be considered as a replacement for mechanical ventilation systems. Think about how often you actually open your windows or how many days each year passive ventilation is a viable option for you. In Chapter 7, we will discuss indoor air quality and how to manage indoor air pollutants in your home. Mechanical ventilation is a necessary component of healthy indoor air quality. More than that, mechanical ventilation is a mandatory requirement in the 2012 International Residential Building and Energy codes. The paradigm is to build them tight and then ventilate them right! We will have more to say on the equipment to use and how to provide ventilation in the different climate zones in our country in Chapter 6. Yes, climate makes a big difference in how ventilation should be done.

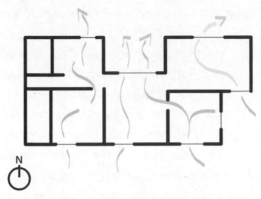

FIGURE 3.4. Cross ventilation.

Cross Ventilation: It is important to design your home with operable windows on the prevailing breeze side of the house (east or south sides, in Figure 3.4) *and* on the opposite side of the house for the purpose of supplying passive ventilation. The prevailing breeze creates pressure differences from one side of the house to the other. Strategically placed windows on the prevailing breeze side of the home and the opposite side work to draw the breeze through the living space. We can actually increase airflow through the space by opening more windows on the negative pressure northwest

FIGURE 3.5. Modern dogtrot. Credit: Frank Ooms Photography.

(downwind/leeward) side than on the positive pressure (southeast/ windward) side of the home. When wind currents travel through a constricted opening, the velocity increases (the Venturi effect). The Bernoulli principle determines that as the air speeds up, its pressure drops, causing it to create suction and drawing in even more air.

The same principle can be applied to the architectural layout of the structure itself, using a vernacular breezeway or dogtrot design to channel breezes into the living space. To increase the breeze's effective cooling, the alignment of the building sections can be offset at an angle. In this way, we can use smaller windows or openings on the windward side of the dogtrot and larger windows or openings in the dogtrot on the leeward side to pull air through the structure. Windows on the side walls of the dogtrot are used to channel that increased volume of air into the living space of each building.

Stack Ventilation: As you may recall from Chapter 1, one of the basic principles we work with is that warm air rises. As this occurs, the area below is left under slightly lower pressure, so there is a suction effect. As the hot or warm air rises, windows located high on the opposite wall, or in an open upper-story area (like a stairwell, clerestory, cupola or monitor), can exhaust warm air out of

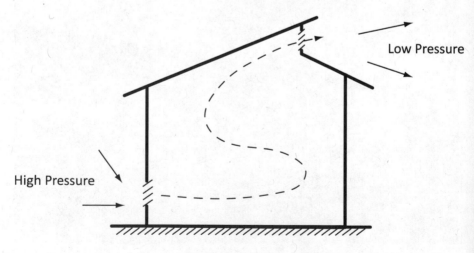

FIGURE 3.6. Stack ventilation.

the house while drawing cool breezes in. Any time we can use these passive ventilation strategies, we increase the velocity, and thus the effectiveness, of even the mildest breeze. Whole-house fans can intensify this effect. These devices are especially useful in areas where summer nights are relatively cool, exhausting warm air that builds up during the day and drawing in cool nighttime air for passive ventilation. If we can provide cool, comfortable air without having to use our mechanical systems, we save energy and money.

Thermal Mass

Before choosing your building system, you should analyze your ability to benefit from the use of high thermal mass building systems in your climate. Thermal mass can be any material that can capture heat and hold it for hours after exposure, then release it slowly. The most common materials used for exterior thermal mass are types of masonry, e.g., stone, rammed earth, adobe or concrete.

During the hot days, as these materials absorb heat, it slowly moves inward through the walls. About the time that the sun sets, the heat has reached the interior space. Throughout the evening, as temperatures drop, the exterior walls cool down because the outdoor temperature is well below the inside temperature, directing the flow of heat toward the outside. The materials continue to release their heat as they become increasingly cooler, until they reach the ambient nighttime low temperature. Then, as day breaks, the interior is able to maintain these cooler temperatures well into the hot afternoon hours as the direction of flow reverses again. This can be one of the most efficient natural heating and cooling strategies available, and effectively replace mechanical heating and cooling systems—for big cost savings!

This is an application of the second law of thermodynamics: heat moves from warm toward cold, and only works if nighttime temperatures where you live drop rapidly to well below the desired indoor temperature and you use very thick thermal mass exterior walls. Building sites with big diurnal temperature swings (high temperatures during the day with starkly contrasting low night temperatures), such as those located in mountainous or arid regions, such as the desert southwest, will benefit the most from using exterior thermal mass.

In areas with long, hot humid summers, where nighttime temperatures don't really cool off that much, exterior thermal mass does not work very well, as the walls never get the chance to cool sufficiently. In this case, the outdoor heat is constantly moving inward, making it impossible for your air conditioner to keep up. These walls do not act like insulation.

If you live in an area with long, cold winters, where it never gets warm during the day, these types of mass walls will constantly move the heat outward. This is one of the reasons why native peoples build lightweight homes in warm, humid parts of the world and heavy-mass earthen-walled homes in arid deserts with hot days and cold nights. They may not have known physics as such, but they figured out what worked through trial and error.

There are other ways to use thermal mass if you don't live in that high-diurnal type of climate. Trombe walls and tile floors abutting south-facing windows that collect solar heat all day release it into the living space at night. Venting the wall assembly correctly to create convection loops intensifies the effect, which can be controlled by closing the vents at night and sizing overhangs for protection in the summer.[4]

In cold climates (and some people even use this strategy in warm climates), we can add thermal mass inside the home as a component of our passive heating design strategy. Concrete foundations that serve as the finished floor (i.e., stained or polished concrete), ceramic tile floors and interior stone walls are the most common thermal mass features used in these climates. To serve this function, the materials must have measurable direct solar exposure for several hours during the day or be used in conjunction with an indoor heating appliance, like a masonry fireplace. Passive solar design strategies incorporate windows and shading devices to control solar exposure to these mass features. It also means that radiant heat is not blocked by any materials specifically designed to do that, such as low solar heat gain coefficient (SHGC)-rated windows. The types of materials specified for each location and application must work together for passive systems to optimize benefits. Again, these strategies can be used to lower your mechanical heating and cooling costs, saving energy and money.

It is important that you understand and build what is right and will work well in your climate. Don't believe that everything you see on TV is right for your climate and will work well where you live. If the building advice or system comes from a cool, rainy and mild climate and you live in a hot and humid climate, beware. Using the perfect applied building science for Minneapolis when building in Atlanta will prove disastrous, and vice versa!

Passive Solar Design

Wherever you live, you can embrace design strategies to further manage the sun's heat loads on the structure. As we mentioned in Chapter 1, in the northern hemisphere the sun is always either in the southern sky or directly overhead. In North America, the sun rises at due east on the spring equinox, April 21, and fall equinox, September 21, and sets at due west on the same dates. On June 21,

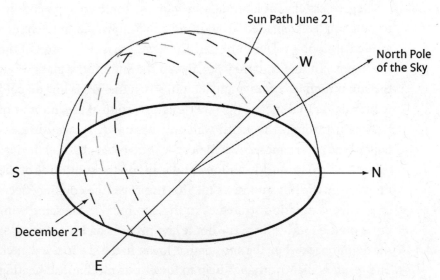

FIGURE 3.7. Seasonal sun paths. Credit: Adapted from Dr. David P. Stern.

the summer solstice (the longest day of the year), the sun crosses at its highest path across the sky overhead, rising and setting well north of due east and west. Each day after June 21, the sun's path starts and ends further toward the south of due east and west and lower toward the southern horizon until it reaches its lowest path across the sky on December 21, the winter solstice (the shortest day of the year), and then the process reverses. Knowing the seasonal sun path across your site allows you to design to manage solar heat loads to your benefit.

Take time to research your location's latitude. The closer you are to the equator (like in the southern US), the more days you will need air conditioning rather than heating. You want to minimize the amount of solar heat gain on the structure, so, as mentioned in Chapter 1, you want the short walls of the house to face east and west; since these two walls have the least amount of wall area, they will absorb the least amount of solar heat into the home.

Interestingly enough, if you live in a predominantly cold climate area like the north (with more days needing heating than air conditioning), you should still design your home with the long walls of the rectangle running east and west.

Regardless of your latitude, you want the south-facing windows to be completely shaded at noon on June 21 and completely unshaded at noon on December 21. The key here is the design of the overhangs. In this case they must keep the sun off the glass when the sun is high in summer and let it in when the sun is low on cold winter days. That means that we design to control the amount of heat gain through the south wall and windows. We can use extended roof overhangs, covered porches and other shading devices (like awnings) to shield windows and walls on the south side of the house from the hot summer sun. This improves the cooling effects of passive ventilation strategies in the summer. These same windows on the south side of the home can provide passive heating of the living space when the sun's path is lower in the sky in the winter. The result is that overhangs help to lower our mechanical heating and cooling costs.

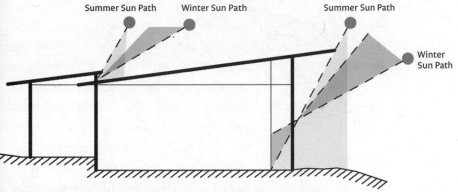

FIGURE 3.8. Overhang depth affects seasonal shading.

As you can see in the drawing above, the lower the sun is in the sky and the shorter the overhang length, the further the sun penetrates into the house. The farther north you are, the lower the sun's path across the sky, and the farther south you are, the higher in the sky it moves. This fact will affect the length of your overhangs and how you build to take advantage of or avoid solar heat gain in your home. All three of these things—length of the overhang, height of the overhang above the window, the angle of the sun in the sky—impact how much solar heat gain you will reap (or keep out, as the need may be).

Shading Devices: Direct sun exposure on walls, especially west-facing walls (and south-facing walls in the summer) can significantly increase the loads on your cooling system—roof overhangs and porch roofs to the rescue! Many types of external shading devices that can keep heat from entering the building are often recommended in lieu of prescribed low solar heat gain coefficient windows on southern exposures in passive solar design. These can include pergolas, awnings and patio covers.

Shading a window on the outside of the glass is twice as effective at reducing heat gain as doing so inside of the house.[5] This is the best strategy for reducing cooling loads and lowering energy costs. Individual awning structures mounted directly above windows can provide more protection, being closer to the window, than an over-

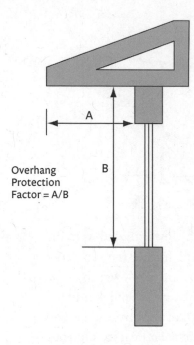

Overhang
Protection
Factor = A/B

FIGURE 3.9. Projection factor.

hang that is several feet above. Of course, the projection factor, or the ratio of the depth of the overhang or awning to the height above the sill of the window below it, will influence its effectiveness in shading all or portions of the window over the cycle of the sun's path between the solstices. By using the projection factor formula, we can design the shading device to precisely control which months the sun is allowed to shine into these windows. Even in the north, you don't need solar heat gain in July.

Never underestimate the benefits of any shading devices. As was previously mentioned, they also serve to protect exterior doors and windows during major rain or storm events. However, if external structural devices are not possible, trees can provide shade, either continually or seasonally (discuss further in Chapter 8). If your site does not have trees, neighboring buildings or other existing features can provide the desired shading on the east and west sides of your home. Other methods of shading are discussed in Chapter 10.

Outdoor Living Space: This is also the time to think about incorporating outdoor living areas into your design. If possible, think about where you would place them in proximity to your home. Before you meet with the architect, take a few stakes and a compass out to your lot and mark off roughly where the house will sit and note where the four compass directions are relative to your home. Think about where the front door will be and where other exterior windows and doors will be. What views do you want to see from your living room and kitchen windows? Now, take a slow walk (with a landscape designer if you need one) around the space just outside of what will be the home's walls. If you are just starting to design the home and you plan to own the property for a while before you build, you may want to take this walk during different times of the day, even during different times of the year. Look at the path of the sun and notice which areas are shady and which have full sun exposure.

In many instances, outdoor living structures can serve as important shading devices for your home, so locating them strategically to serve this purpose is something that should be integrated into your overall passive design strategy. This means that these spaces should be connected to your house with some sort of shade cover, pergola or roof. The best place to locate it is on the side of the house that benefits from the prevailing breeze. Think about how you can capture and direct the breeze into the space you have selected.

If you plan to build in a development where houses are close together, you should consider a usable front porch, facing the street, to provide opportunities to interact with your neighbors and build a sense of community, developing the relationships that will benefit all in the long run.

In a cooling-dominated climate, plan the location of your summer kitchen on the east or north sides of the house, as these have the least direct solar heat gain during the summer months when you would most be using that space. These areas are usually the shadiest, since the home itself will provide shade most of the year in the afternoons and evenings when you would be most likely to cook outdoors. Also, you want that area to be in the direct path of the prevailing breeze. The addition of a ceiling fan can extend the number of months of enjoyment, so think about whether your design needs to include electrical outlets.

Also, think about how you might be able to use the space at night as well as during the day—perhaps a place surrounded by trees that you can hang some solar lights from and have for outdoor dining. Think about plumbing for an outdoor sink and shower and whether you want these areas covered so you can use them during wet weather.

Next, if you live in a predominantly cold climate, if possible plan the outdoor space on the west or south side to lengthen your outdoor season. Having a covered four-season porch on the south side of your home will give you outdoor space that you can use even in the winter months, when the sun is lower on the horizon in the southern sky. The sun's angle will help keep that space warm, and you can enjoy reading a book, taking a nap or just sitting and enjoying the view and breathing some fresh air. There's no better way to beat the winter blahs than getting outside for a bit of sun.

Think about how you might enjoy your outdoor space in the winter months. In hot climates, indoor fireplaces cause more heat loss problems than benefits. This not only wastes energy for heating, but can also create negative pressure in a house, so that cold outdoor air is being sucked in from any building crevice available. Moving the fireplace to an outdoor location (maybe on that screened porch?) can help remedy that. If you locate this in an open space make sure that is far enough away from trees and vegetation to prevent a fire hazard. More details on outdoor living are covered in Chapter 8.

Designing for Natural Daylight

Daylighting serves as a passive lighting system, and natural light can significantly improve our mood (and treat Seasonal Affective Disorder, or SAD). Daylighting is most effective as a passive feature when it provides enough lighting for you to function around the house without needing to turn on electrical lighting during daylight hours. You should be able to move from room to room, navigate through a room or do things that do not require detailed vision without having to turn on any electric light fixtures. This includes using the bathroom, opening the refrigerator, placing dirty dishes in the dishwasher, etc. This will save energy use and, therefore, save you money.

Do not assume that all windows provide daylighting. We have walked many new construction inspections with homebuilders claiming daylighting but with only standard windows installed. Prior to occupancy, those windows are bare and allow maximum daylight to enter the space. However, in most instances, once the home is occupied, those windows will be covered with blinds or draperies for privacy. More often than not, to function within those spaces, those same occupants will not walk over to the window to open the blinds or draperies, they will just turn on a light switch. Consequently, standard windows that have a reasonable expectation of being dressed by privacy treatments do not qualify for daylighting.

Clerestory and transom windows, being high up on walls, throw light across the ceiling and provide general illumination, but are high enough on the walls that you do not feel the need to install any kind of privacy treatment, like blinds or drapes, on them. These windows are so high on the wall, anyone outside of your home would need a ladder to see into the home through them, even at night. However, if you are building next door to a two-story home, even windows high on walls may not provide privacy from your neighbors' upstairs views. If this is the case, it's best to save your construction budget for other energy-saving upgrades, as securing your privacy will negate the intended daylighting benefits.

In warm climates, transom or clerestory windows, because they are high on the wall, should be placed under shading devices (roof overhangs, porches or awnings) so that they are protected from direct sun and water intrusion. If you are installing a cupola or monitor or even using transom, clerestory or double-hung windows to exhaust heat, these windows provide double duty by offering daylighting and passive stack ventilation. Of course, the windows must be operable in order to serve this function.

However, you will need to install some light control materials on daylighting windows in sleeping rooms. When you're ill, on your day off work or want to sleep late on the weekend, you don't want to be disturbed by daylight entering the room. Interior shutters work well on bedroom windows. When you get out of bed, just open the shutters instead of turning the light on.

Skylights: Depending on your climate zone, skylights might or might not be appropriate. In hot southern climates, where the sun's position in the sky is more directly overhead and the roof of the structure is dealing with the brunt of the heat gain in the hot summer months, skylights are usually best avoided. The exception to this is if they are ENERGY STAR-rated, and even then, applications should be limited. We know some manufacturers include tinting or integral blinds that can minimize exposure, but this might not be enough to offset heat gain through this opening. After all, by the time the heat has hit the blinds, unless the blinds are between the panes of the glass, the heat is already inside the structure.

Even in milder or northern climates, you need to remember that every penetration through the roof is a potential water leak. Skylights, like other windows, will lose their seal eventually and leak, requiring replacement. Other strategies for daylighting and passive solar can achieve better results than skylights.

For daylighting in rooms or interior hallways without a window, consider installing sun tube devices, which bring in light from the roof without bringing in the heat. They utilize a flexible tube that functions like a big fiber optic cable and is installed through the attic. Or consider installing interior windows or openings in walls to allow light into areas that do not have direct access to an exterior wall.

Floor Plan Room Layout

Consider the kitchen as a supplementary heating device for your home. In hot climates, it's typically better to place the kitchen on the side of the house opposite the direction of the prevailing winds. If the prevailing wind is out of the southeast, place the kitchen on the north side of the house. This will help exhaust heat out of the house in the summer. In the winter, north winds will push heat from cooking appliances back into the living space. Even in predominately cold climates, a north kitchen location is preferred, as it effectively doubles the return on investment on whatever type of energy is used for cooking.

The same benefit applies to heat generated from your laundry room, especially if you use your clothes dryer. Again, locating it on the north side of the house with the dryer connection on the exterior wall might reduce the loads affecting your cooling system. The best practice would be to locate the laundry room outside of the conditioned living space, to keep the moisture and heat loads outside. If there is a large, open unshaded area on the west side of the house, plan to use that space for a clothesline.

Also, consider how room placement could provide additional insulation of main living areas. In hot climates, try to flank your west-facing walls with closets, garages, utility room or other buffer space to mitigate direct afternoon sun into your living space. These buffer spaces will reduce the heat infiltration into the living spaces, lowering your cooling loads and cost of operations.

In a cooling-dominated climate, the west wall should be protected by deep overhangs, porches, awnings or seasonal shading devices like trees or vines. Better yet, build a detached garage or carport on the west side, which will still work to shade the building while separating pollutants like auto exhaust and any chemical products stored in the garage away from the living space of the home. Of course, in cold climates you want your main living areas on the south side, for passive solar gains. The south side of the house would be a good location for a solarium or sun room and some type of thermal mass like a tile floor, as these south-facing glass-enclosed rooms provide additional winter sun exposure in cold climates. Sunrooms are almost never appropriate additions in hot climates.

For raising a family, consider having all the sleeping areas upstairs. Locating all bedrooms upstairs allows us to install a zoned HVAC (Heating, Ventilation, Air Conditioning) system, which means separate control thermostats for each floor. This allows for managing upstairs comfort in sleeping areas at night separately from downstairs comfort for living areas during occupied, active hours, reducing operating costs and extending the life of systems.

There's no reason to have a master bedroom downstairs unless you expect an aging relative to live with you while you are raising your family. Once their kids are grown, most couples downsize to a home that is more economical to maintain. The majority of seniors prefer single-story homes, as climbing the stairs can become increasingly more difficult in aging-in-place scenarios. As we discussed in Chapter 2, a better alternative is to design your home to accommodate your changing needs over time. This "convertible home" is discussed in more detail in Chapter 15.

Sleeping rooms were popular in warm climates before we had air conditioning. Screened porches were on the side of the home that could capture the prevailing breeze. We can still locate our bedrooms on this side of the house and open our windows at night during milder seasons, to sleep comfortably without air conditioning. This extends the non-cooling season, saving energy and money.

3. Designing for Resource Efficiency
Basic House Design

Typical homes built to current standards do not generally make it more than twenty years without needing some repairs. Many factors contribute to this, but certainly any complexity of the design,

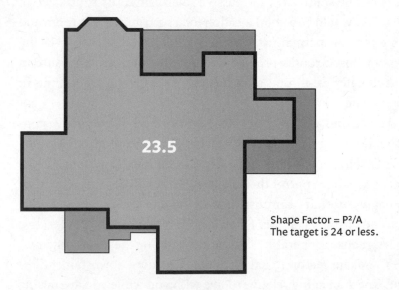

23.5

Shape Factor = P²/A
The target is 24 or less.

FIGURE 3.10. Shape factor.

creating potential for water leaks over time, is a significant influence. To reduce this risk, it's best to keep the basic house design as simple as possible. Starting with the foundation, try to *stay as close to a true rectangle as possible*, avoiding projections in the footprint. Starting here, everything above the foundation then follows suit, you'll have a simpler basic structural design, and a simpler roof design will follow. All of these efforts result in lowering the base bids for construction costs. This is one of the best strategies for leaving room in your budget to fund high-performance green upgrades discussed in Chapters 4 and 6. If you want the "green bling" we discuss in Chapter 9, you need to keep the basics simple so you won't run out of money before you get there.

The shape factor concept used in designing post-tension foundations (SF=Foundation Perimeter^2/Foundation Area)[6] indicates that changes in direction or projections require foundation engineering and reinforcements that not only add cost to the foundation, but also increase the incidence of failures. Beyond concerns for the foundation itself, all deviations from a basic rectangular design add complications to structural assemblies and roof designs that are not only costly to build but again have greater potential for leaks. Costs increase not only because of the additional materials required to provide structural support, but also from increased labor that adds time and costs to construction schedules, impacting interim financing and construction costs. And it's not just the money—chopped-up architectural designs waste natural resources, as they take much more materials to build the same basic square footage.

We also found a study that references shape factor in terms of the relationship between a building's thermal envelope area and the volume of its conditioned space. From this perspective, buildings with higher shape factors, i.e., a larger area of exposed thermal envelope, use more energy than those of the same area but a smaller shape factor.[7] This was especially true in cold climates due to an increased area of heat loss, and had less to no impact in climates as they progressed along a continuum to warmer.

Designing for Two-Foot Modules: If you really want to optimize resources, create the least amount of waste and cut your base construction costs, design the building on two-foot increments. Most building materials come in two-foot incremental lengths or widths. Most lumber and millwork come in 8-, 10-, 12- and 16-foot lengths and sheathing, decking and drywall materials in 4×8, 4×9 or 4×10 sheets. Designing for two-foot modules is one aspect of Advanced Framing, which is a more efficient way to frame from both a materials use and energy use perspective (more on this in Chapter 4). So, overall exterior wall dimensions on your plans should be exact two-foot increments, and plate (ceiling) heights should be 8 feet, 9 feet or 10 feet (there is no reason to go over 10 feet) for optimal material efficiency and cost.

The same is true for interior room dimensions, but more with respect to finished flooring materials. Sheet goods (carpet and linoleum) come in sheet widths of 6 feet, 12 feet or 15 feet, and tile and wood products typically come 25 square feet to a box. If you are not designing to those increments, you are going to waste a lot of materials. The worst case would be to design a room 12 feet and 7 inches wide or 15 feet and 2 inches long. If you are specifying sheet flooring, your installer will need to cut a 7-inch-wide piece to finish off the 12-foot standard roll width. Or for boxed goods, cut 2-inch strips to finish off that 15-foot full square on each run. Since some wood flooring is tongue-and-groove, once the two opposite edges of the piece are used to fit the adjoining pieces, the rest of the square is waste as there are no usable edges remaining, or will require additional labor to router onsite.

Your building designer should hand you the finished plan sets, and you should easily see that overall building dimensions (not including exterior cladding) are measured in even 2-foot increments. This means that from corner to corner, sole plates, top plates, stud spacing, etc. all fit within the standard measurements that wood framing is manufactured to, so no waste. This is the most efficient use of your money, as there is nothing worse than wasting it on material scraps in the dumpster!

Stacked Stories: Foundation, wall and roof assemblies represent large portions of typical construction budgets. Stacked structures with more than one story can save considerably on material resources and therefore cost. Using in-line framing techniques (another aspect of Advanced Framing discussed in Chapter 4), the floor, wall and roof framing members are vertically aligned with one another so that loads are transferred directly downward, providing

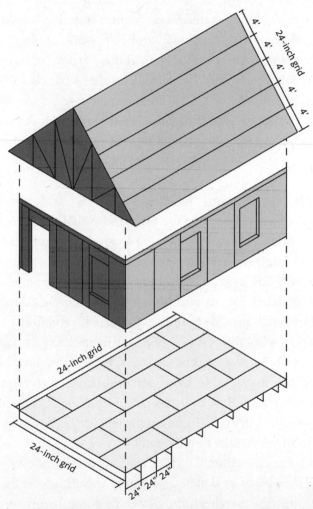

FIGURE 3.11. Stacked stories. Credit: Adapted from Office of Building Technology, State and Community Programs, Energy Efficiency and Renewable Energy, US Department of Energy.

a sound structural assembly using less framing lumber. Also, since multi-story homes have a smaller footprint than single-story structures, they can be built at higher density, reducing land costs for affordable urban sites.

Wall Design: To optimize your design for resource efficiency, you will need to decide what type of structural system you will employ. As we discussed in the section on "Designing on Two-Foot Modules," various building materials are manufactured in standardized sizes. In addition to dimensional lumber, this also applies to many alternative and natural building materials. Straw bale comes in standard bale dimensions, structural insulated panels (SIPs) and insulated concrete forms (ICFs) come in standard 4-foot wide and 8-, 9- or 10-foot heights, and autoclaved aerated blocks come in fixed block sizes. Knowing which type of system you will use will save you money on both materials and labor.

Framing Details: Framing details are necessary with almost any building structural system in use today. Straw bale buildings must have structural support members for load-bearing walls and roof assemblies, rammed earth buildings need structural framing for door and window openings, and of course, wood-framed homes need lots of details for structure and weatherproofing assemblies.

Design details provide templates for structural stability to keep the roof on the building (very important!), especially during high-risk events like snow storms or tornados. Wall assemblies also keep air infiltration manageable, help keep rain-driven water out of the wall assembly and allow moisture and water vapor that does get inside the wall to dry out. These details will vary depending on the climate location that the home is built in.

Your home plans should include framing details for every building section, as well as detailed assembly specifications for walls, foundation, roof members and water flashing details for roofs, walls, windows and doors. This book talks more about building materials and systems in Chapter 4, but basic principles related to the building envelope are more influenced by design than the materials themselves, so they are best addressed here. By providing framing details in the drawings, framing contractors onsite are less likely to over-build assemblies, thinking that it is better to err on the side of adding more lumber than is necessary for structural integrity than not enough. Do not assume that the education level of a framing crew provides expertise in structural engineering or waterproofing. Instead, have your structural engineer or architect provide the framing and waterproofing details for wall, roof and floor assemblies (if you are not using engineered trusses or panelized assemblies). This will save you money because every material you purchase will be used as it was intended, instead of being misused, requiring you to purchase more materials either during construction or to repair poor workmanship later. Often construction drawings give the crew no clue as to how to waterproof and flash the critical openings and junctions in the house where most leaks will occur. A recent article in *Builder Magazine*[8] found that about one-third of US homebuilders still don't properly flash windows to prevent direct intrusion by rain. It stated that water leaks, especially around windows due to improper flashing, "routinely tops the list of builder callbacks." It is imperative to the durability of your home that the flashing materials and their installation be detailed on the drawings.

Good building science is the perfect partner to green materials and design. Using the right materials to create the right building system for your climate will maximize the durability, healthfulness, comfort and efficiency of your new home. Make certain that your architect or engineer is on-board with your desire to maximize structural integrity while reducing waste and achieving a watertight, durable and healthy structure.

If you have hired a custom builder for your project and are not working with the architect directly, you should make certain that the home plans that the builder is offering you have been designed according to the strategies outlined in this chapter to ensure that construction will be resource efficient. You should also ask the builder for written specifications for the basic building materials that are included in his quote for the home. Better to find out up-front if there are any surprises coming down that will throw you off budget.

Designing for Durability

The best practices in building materials and methodology for each type of climate have been vigilantly researched and tested by reputable groups like the Building Science Corporation and the Building America Program run by the US Department of Energy's Office of Energy Efficiency and Renewable Energy (www1.eere.energy.gov). Joseph Lstiburek, PE, PhD and Betsy Pettit, FAIA, who head that firm, have described these practices in a series of books, entitled *Builders Guide for…Climates*, which includes books detailing appropriate building science strategies for hot and humid climates, mixed-humid climates, hot-dry and mixed-dry climates and, of course, cold climates. These building guides include hundreds of detail drawings, various options for each assembly and an explanation of how and why the system presented is right for each climate. This can be very useful when it comes time to convince your builder, architect or subcontractors to change their old-school ways. These books are available at the Building Science Corporation bookstore (buildingsciencepress.com/Builders-Guides-C1.aspx).

Dr. Lstiburek has contributed to countless research projects leading to the development of building science best practices and influencing many building codes and standards. We consulted one of his publications, "Read This Before You Design, Build or Renovate,"[9] numerous times in writing this book.

You should check with your local building inspection office or other agencies that can provide you with building codes applicable to your area, as well as any specific wind or seismic ratings required for building materials. Depending on risk factors for certain natural disasters, you may be required to use particular assemblies or materials in your home's construction. Some common risks were discussed in Chapter 1. Building materials are available that are termite

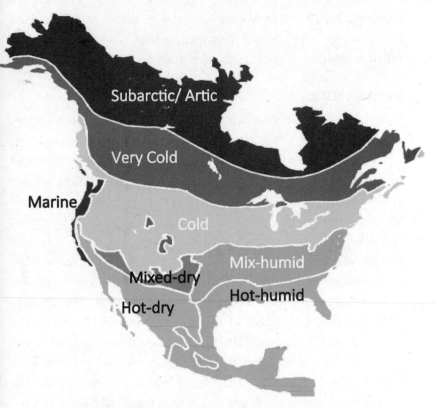

FIGURE 3.12. EEBA Climate zone map. Credit: Adapted from
Energy and Environmental Building Association.

resistant, or not termite edible, or able to withstand storm events.
For example, windows may be required to meet a certain design
pressure (DP) rating in order to withstand high-velocity winds
from tornados or hurricanes, or sheathing may be required to be
nailed up in a particular pattern to meet seismic load requirements.

Durability is also a function of pressure differences. If you are
building in an area that has frequent exposure to wind, sun or rain
loads from a certain direction, your design and construction details
should include strategies to mitigate long-term damages from these
forces on the structure. Provide drainage to protect the foundation
from uneven drying after a rain event, protect windows and wall
assemblies from wind-driven rain or constant sun exposure and
add structural framing to balance wind pressure or snow loads, as

required. Building right to begin with means that over the life of the home you will save yourself from wasting money and efforts repairing damages that resulted from predictable risks for your location.

Mechanical Design

Designing for resource efficiency involves more than just designing the building's structural components. Often mechanical system design and the space required for locating the equipment and air ducts within the structural components of the building are not given any consideration by the architect. If you have any experience with mechanical, electrical and plumbing (MEP) plan review, you will see that our current methods of designing these systems is archaic. More often than not, mechanical contractors are forced to work with structural plans that do not provide an efficient path for system component installation. As a result, although their designs reflect the size and length of service runs required based on the room layout, the actual installation must be adjusted around obstacles in the structure. To compound that, most field personnel are poorly trained in understanding how to effectively install materials for efficient operations of systems. These issues add costs to your initial construction bids and to your operating costs over the life of the home and explains why so many homes operate inefficiently and have comfort challenges.

Do not go any further in your design without thinking about the location of the main systems that you plan to install in your home, especially expensive, high-efficiency-rated equipment. Plan and design the home with the space needed to locate this equipment in the center of the spaces that they will serve. This applies primarily to both the water heater and the air conditioner/furnace blower. This is the part of the HVAC system that moves hot air from the furnace and/or cold air from the air conditioner to all the conditioned living space. Centrally locating these systems shortens plumbing and duct runs, so that air and water are moved most efficiently with the least energy losses over distance, saving you money both on your initial installation bid and on long-term utility costs.

Plumbing Design

With an exploding world population and shrinking supply of fresh water on the planet, many say (and we agree) that water resources are the next gold standard. In many areas of the country, potable water costs have doubled and tripled in just the last couple of years.[10] To survive the coming supply and demand problems, we are going to have to find ways to get the most from every drop.

Plumbing piping systems provide both hot and cold water delivery, so design considerations provide the greatest benefit for indoor water conservation. It is important to remember that water stays in the plumbing lines all of the time. So every time you turn on the hot water, you will be wasting all the water already in the line waiting for that hot water to get to you from the water heater. This means you are paying for and wasting water that you never even use!

Efficient hot water delivery is achieved through a whole-system design approach, not just considering the type of water heater to be installed. Nothing can kill the efficiency of a high-efficiency water heater more quickly than a poorly designed distribution system. For efficient hot water delivery, you must have one of the following:[11]

- Water heating equipment located within 20 feet (30 feet on 2-story homes) of all the wet locations in the house. This could be a tanked or a tankless unit.
- A manifold plumbing system (replaces the main water line with multiple smaller lines run to each fixture) with the manifold (split) centrally located within 10 feet of the water heater and no single branch line length exceeding 15 feet.
- A structured plumbing system (not a manifold system) consists of a main trunk line that services the entire house with smaller branch and twig lines to each wet area and fixtures, *with* an on-demand water recirculating system installed at the fixture(s) farthest from the water heater, *and* pump controls located at all wet locations along those long runs in the home.

Another critical element in achieving efficient hot water delivery is pipe insulation. No matter how efficient your design is in delivering hot water to the fixtures, we still see significant temperature drops during transmission or while water is sitting in the lines waiting for the next draw to occur. Also, when the hot water pipes are not insulated adequately, there are substantial heat losses from hot water running through pipes within the conditioned space of the home that add to the cooling load of the air conditioner. Insulation is cheap, easily installed and means hotter water at the tap for you and your family with a lower temperature setting on the water heater. Insulating three-quarter-inch pipe with three-quarter-inch insulation will triple the time before the water cools down; insulating half-inch pipe with half-inch insulation will double the cool-down period.

Central Core Plumbing Design: A large percentage of the cost of plumbing (both materials and labor) is due to long pipe runs. These same long pipe runs also mean more line full of water all the time, waiting to be wasted down the drain whenever you turn on the hot water tap. Shortening the plumbing runs saves initials costs of plumbing the home and saves water costs for the life of the home.

The most efficient plumbing design is to locate all of your wet areas (kitchen, baths and laundry) adjacent, or at least close, to each other and the water heater. If you are building more than a one-story home, the same is true for stacked plumbing locations, where upstairs bathrooms and laundry rooms are stacked directly above downstairs kitchen, laundry and powder room locations. This will deliver the best hot water service with the least amount of waste. Even if your home cannot be designed with a compact plumbing scheme, try to centrally locate the water heater at or near most of the wet areas.

Central Manifold System—Beware of What you Get! Many plumbers believe that installing the manifold centrally gives you a little more flexibility on where to locate the water heater tank, while still managing water delivery from a centrally located manifold. However, central manifold systems,[12] if not limited in run length, can be the most wasteful in terms of both water and energy, even less efficient than daisy-chained piping that runs the entire run around the house. But if you just cannot find a way to locate all your wet locations near each other or centrally locate your water heater to service wet areas within a short distance of each other, then you might consider a *well-designed* manifold system.

It's important here to reiterate that water stays in the lines at all times. The piping from the water heater to the manifold goes cold about 5 to 22 minutes after using the hot water (in uninsulated three-quarter-inch pipe, less time in smaller pipe), as does the water in the branch lines to each fixture. Therefore, central manifold systems are only efficient if the manifold is plumbed the shortest distance possible from the water heater and the wet locations are also short pipe runs from the manifold. This, again, means short *overall* distance from the water heating appliance to each fixture.

To illustrate this, imagine that your master shower is 40 feet (we mean 40 feet of pipe, which given the way it's routed and goes between floors might look like 20 feet in a direct line) from the water heater (which is located in the garage), and the kitchen is 30 feet in the same direction, but the secondary bath is 20 feet in another direction. If you installed the manifold close to the kitchen (the highest hot water use location), the distance from the water heater to the second bath increases while the distance to the kitchen and master bath is reduced. When you shower in your master bath, all the water in the main line to the manifold and in the branch line from the manifold to the master bath is lost down the drain waiting for hot water delivery. If you get out of the shower and walk over to the master lavatory to shave, where is the hot water? At the manifold (remember, each fixture is run directly from the manifold). So, again, you are wasting all the water down the sink drain that was in the branch line between the manifold and sink. If the house had daisy-chain or structured plumbing, the hot water would have been at the shower and would have only had to travel a few feet in the line to the vanity sink.

If you want to cook breakfast after showering, where is the hot water? Well, if you can make it from the shower to the kitchen within 10 to 15 minutes, the hot water is still at the manifold, and since you located the manifold close to the kitchen, you'll get hot water pretty fast. But what are the chances that you won't make it to the kitchen within 10 to 15 minutes of getting out of the shower? It's more probable that the water in the line will have cooled down and it's a cold start from the water heater again, and all of the water in the main line to the manifold and in the branch line from the manifold to the kitchen is lost down the drain waiting for hot water delivery.

What if your child wants to take a shower right after you have showered? Where is the hot water? Again, at the manifold. But we increased the line length to the second bath when we installed the manifold near the kitchen, so again, much water is wasted down the drain because the branch line from the manifold to the second bath is full of cold water. And if your child waits more than 10 or 15 minutes after your shower to start their shower, they are back to a cold start all the way back to the water heater.

So unless you precisely time all of your hot water events to occur consecutively, you could actually be worse off with a manifold system. In this case, it ends up being a very inefficient and inconvenient plumbing design, except where the manifold consists of a short trunk (manifold trunk line runs a maximum of ten feet from the water heater) and branch lines (maximum of 15 feet to each fixture).

Structured Plumbing:[13] A greater water savings can be realized by downsizing the diameter of the pipes and reducing the number of runs using the structured plumbing approach. With today's more water-efficient fixtures and appliances, we no longer need three-quarter-inch runs or even half-inch runs to each outlet. This not only saves on your plumbing materials, it also saves water over the life of the home.

The structured plumbing design uses twig lines that typically extend no further than ten feet from the main trunk, with the ex-

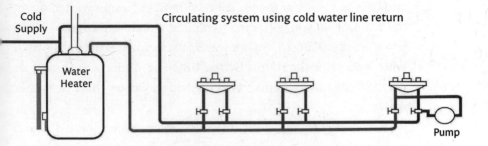

FIGURE 3.13. Structured plumbing using on-demand recirculating pump. Credit: Gary Klein, Affiliated International Management, LLC.

ception of large volume fixtures at garden tubs, clothes washers or sinks in a kitchen island that run the water line through a concrete slab. This type of system uses less pipe and can be installed faster, saving money on your base bids. Since water stays hot longer in the larger main trunk line, only the water in the twig lines is lost down the drain when hot water is already in the main loop.

The best strategy for reducing water wasted down the drain waiting for hot water delivery is a structured plumbing system controlled by an on-demand pump. This can use a closed hot water circulation loop with a dedicated return line to the water heater or use the cold water line to return water until hot water is delivered to the open fixture.

Even if your home design is plumbed through using a standard plumbing branch and twig run, you can still install an on-demand recirculation pump at the fixture farthest from the water heater. If your plumbing is designed such that you have numerous wet locations in different directions some distance from the water heater, you may need to install multiple demand pumps at the furthest fixture on each run. The pumps are operated by a wireless doorbell button or light switch that is activated only when hot water is needed at that location. Other wet areas along the same line may also have activation switches to run the same pump at the end of the line. The pump contains a temperature sensor and will recirculate the cold water back through the line until the set temperature is reached. When you hear the pump cut off, you can open the faucet and have hot water at the location without any water wasted down the drain. Soon you will learn to flip the switch or push the button if you are preparing to use hot water, just as you flip the light switch when you enter the room. By the time you undress to shower or get the dinner ingredients out of the fridge, you will have hot water delivery.

Resist suggestions to install a motion sensor on these demand pumps. Think about how often you go in the bathroom or kitchen and do not need hot water. If you have motion sensors installed, the pump will run every time, using energy even when you don't need it. And don't even think about installing a continuously operating circulating system, even if you intend to put it on a timer. Consider how many minutes you actually use hot water in the morning or evening compared to the number of hours you would set the timer to run constant circulation. Your water usage will vary day to day, even during the work week when you are on a more rigid schedule. And what about weekends? Are you going to set the timer to run from 7 AM to 11 PM because you do not know when you might need hot water with everyone in and out of the house all day? This is very wasteful and expensive to do.

These on-demand hot water systems pay for themselves quickly when the dollars saved in both lower water bills and reduced water-heating costs are combined. The added convenience of not having to wait standing outside the shower watching water go down the drain is a great plus too. Compared to a constant circulation hot water system or a tankless water heater at each wet location (your only other options if you want hot water fast), these on-demand systems use far less energy since they only operate a few minutes per day instead of constantly. Convenience, water conservation and lower utility cost make this the best option to consider if you cannot design a central core system.

Water and Wastewater Reuse

Now let's take a minute to think about stormwater, wastewater and sewage. These run downhill and offer opportunities for reuse that add value and reduce costs over time. Make certain you plan accordingly.

- You should plan to install a gutter system to at least divert stormwater a minimum of 5 feet from your foundation, to help keep your slab or basement dry and prevent structural damage as well as erosion. It is also a good idea to ensure that the soil slopes away from the foundation at a rate of six inches of drop for every ten feet of distance in every direction. However, getting the water away from your foundations does not mean it goes to waste. We'll cover ways to capture and use this water in Chapter 8.

- If you plan to collect rainwater, think about the roof design and how the gutter system will work to collect the most water from each rain event and get it efficiently to a storage tank. Potable rainwater systems require complex filtration and treatment systems, so be sure to include space for these in your design, if such collection is permitted in your area. Note that six-inch-wide gutters are recommended for this type of rainwater collection. Graywater systems are allowed in some jurisdictions and reuse water from laundry, lavatories and showers/tubs for landscape irrigation (but not for any vegetation that produces edible parts for human consumption).

- Septic systems treat blackwater from toilets and discharge it by methods determined by state and local water quality regulatory agencies. Some types of systems provide irrigation to non-edible landscaping. Composting toilets provide rich nutrients for soil amendments (again, for non-edible landscaping).

You should verify which, if any, of these systems are allowed in your area. If so, you should consider their value in water savings and utility cost savings (yes, part of your water bill is for wastewater and sewage treatment). These systems will be covered in more depth in Chapter 6. You should also check with your water utility to determine if future policy changes might increase the value of those investments. As our water resources continue to be challenged, more creative solutions to conservation are being recognized.

Heating and Cooling System Design

Most heating, ventilating and air conditioning (HVAC) contractors are unaware of the systems approach used in building science and green building. For this reason, they often oversize equipment because they fail to recognize the inherent efficiencies achieved when a whole-house design approach is employed. Even if they are aware of good design principles, for the reasons previously stated regarding inefficiencies in home design, they can't employ them on the job so they just oversize the equipment and hope that extra capacity can overcome installation issues. The result is systems that are more expensive to install and less efficient to operate.

All HVAC systems should be sized and designed using the Air Conditioning Contractors of America (ACCA) Manual J 8th Edition or later software, the ducts should be sized to ACCA Manual D and the equipment selected to meet ACCA Manual S. Properly sizing your system is the key to purchasing the right system to begin with and to efficient, comfortable operations. This is not only wise and efficient, it is now a requirement of our national building codes.

As mentioned earlier, the air handler (what most people call the furnace; it houses the fan for both of the systems) should be located in the center of the conditioned living area that it will service. This location provides the shortest pipe and duct runs. Conditioned air losses due to duct leaks over long runs contribute greatly to inefficiency, regardless of how efficient the equipment you install is rated. So, think about where to centrally locate the air handler so that each run of ductwork is as short and well sealed as possible. National building codes now require all duct systems to be tested and verified as airtight. Be sure your ducts are tested and that they pass code requirements. In many areas of the country, ducts still leak an average of 30 percent of your expensive conditioned air to the great outdoors. That is very costly, as you can imagine, and leads to big comfort complaints, too!

It is also very important to think about and plan routes through the framing members (floor trusses and vertical chases) so that the ducts can be installed with as few twists, turns and compressions as possible to get air to each room or area of the home. If the duct design is done before construction begins, as it should be, the contractor can provide this layout to the truss company. The truss maker can then easily create large square openings in the trusses with the computer software used today to design and build engineered trusses. Often this can be done without an added charge.

If you plan to insulate at the attic floor, consider having the roof trusses designed like floor trusses at the base of each truss, so that the ductwork can run through the lower part of the truss web, inside the area that will be insulated. Or it might be necessary to drop the ceiling in areas of the home or frame a furrdown (enclose a boxed chase along the perimeter of the ceiling) to keep the ducts inside the thermal envelope. Chases, the chambers that house ducts running vertically between floors, should be kept within the thermal envelope by installing plywood, OSB or drywall and sealing and caulking them at the attic floor level (for a vented attic insulated at the attic floor). If these are located on an exterior wall, this means insulating and installing an air barrier on the inside of the outside wall within the chase, since that wall will not be covered by drywall. For horizontal duct runs, try to avoid running ducts all the way out to the exterior walls of the home as was once common practice. It's best to terminate duct runs at an inside wall of each room or in the ceiling near an inside wall. This is a more efficient duct design model that has been created to take advantage of the better windows and insulated walls in our homes today. The US Department of Energy has termed this "compact duct design,"[14] as opposed to the old way of running ducts out to near the exterior wall and then trying to throw the air back toward the center of the room or to wash the wall with conditioned air. The shorter duct run will reduce the initial cost of the system, and the reduced distance that the air has to travel will improve system performance and overall comfort.

Finally, make certain that your supply and return registers are working to support good airflow. Return air grilles should be sized according to the amount of total cubic feet per minute (CFM) of air being delivered to the space. Filter return grilles are (by ACCA Manual D criteria) sized at 200 CFM per gross square foot of grille, and open return air grilles are sized at 300 CFM per gross square foot of grille. Return grilles are very often undersized, leading to high bills, noisy registers and poor comfort. Supply grilles should be the curved-blade type. These will do a good job of distributing the air long distances to cause circulation, mixing it with the stale air before reaching the return air grille. Inexpensive stamped metal supply grills do a poor job of delivering the comfort that your system was designed and installed for. This is not the place to save money.

The term "pressure relief" refers to the ability to equalize pressure between rooms with closed doors, where air is delivered through supply vents for the HVAC system but has no way to get

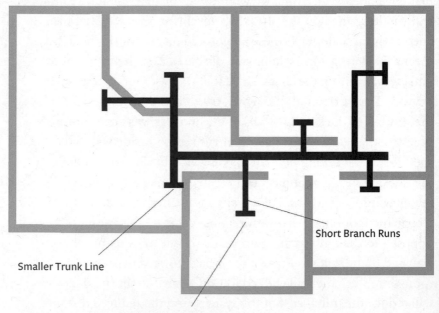

Smaller Trunk Line

Short Branch Runs

Registers Located Near Trunk Line

FIGURE 3.14. Compact duct design. Credit: Adapted from ENERGY STAR.

back to a return air grille to be circulated. If doors are closed and some significant portion of the delivered air cannot find its way back to the return, those rooms will be positively pressured, resulting in other rooms in the home experiencing simultaneous negative pressure because the air handler is trying to find the missing air that it sent out through the supply ducts.[15] This negative pressure causes the system to find other sources of makeup air to meet its capacity. It's possible to draw unwanted contaminated air from the path of least resistance, including combustion appliance vent pipes, fireplace chimneys or adjacent unconditioned spaces (attics and crawlspaces). This introduces contaminants affecting indoor air quality (more on this in Chapter 7).

This means that you should provide a means for the air supplied to a room to get back to a return air grille. This can be accomplished by placing jumper ducts (ducts that "jump" overhead from the enclosed room to an adjacent open space, like a hallway), transfer grilles (same concept as jumper ducts, only the grilles are through the wall) or dedicated returns (each room having its own return air back to the air handler of the HVAC system). All of these strategies

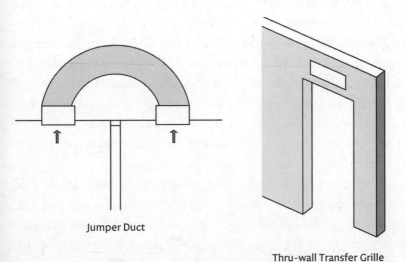

Jumper Duct

Thru-wall Transfer Grille

FIGURE 3.15. Pressure relief. Credit: Adapted from US Dept. of Energy, Energy Efficiency and Renewable Energy.

add up to big savings on efficient operations over the life of your home as well as improved comfort.

Electrical Design

Few residential projects that we have seen to date have given much thought to energy efficiencies when it comes to electrical design. Wiring for lighting is mostly a switch on a fixture, whether it's your dining room chandelier or the walk-in closet fluorescent. If you go out to your breaker box and start flipping breakers, you might find that closet light fixture is on the same circuit as several wall plugs, one of which your plasma television and cable box are plugged in to. Then there's the daisy chain of recessed can lights in the kitchen tied to wall switches at room entrances and exits that just happens to be on the same circuit as the plug for the undercabinet light above your workspace. That doesn't sound like a scheme that supports any kind of control for energy efficiency, does it?

If we want to design and manage energy effectively with electrical controls, we need to first understand the concept of critical loads. Home-run wiring (or "structured wiring") really started gaining market share in the 1990s for use in home security systems and lighting controls. This same concept of managing interior electrical wiring can give us the ability to say, "This circuit has the refrigerator on it and should be operated by the backup generator if we have a power outage" or "This circuit has the electronic devices for the entertainment system and it can be shut off completely between 8 AM and 4 PM, when nobody is at home, Monday to Friday." In other words, this provides us with opportunities to put things you want to control on circuits that you can control.

As we continue to see large improvements in the efficiency of our major energy and water heating systems and equipment, the remaining part of the pie, "lighting, appliance and miscellaneous electric loads" or LAMEL, represents a larger percentage of the load that spins your electric meter. As energy costs continue to rise, the ability to manage this usage will be imperative. Before we go further, we want to note some other terminology here. The terms "phantom" and "vampire" loads both refer to the same 24/7 draw of energy required to keep the power light lit on your plasma TV, the cable box and the clock on your oven, as well as a myriad of other electricity-dependent devices, including anything that has a little box called a converter or transformer plug-in device on it. These converters and devices draw electricity all the time, even when the appliances are not in use, even when our cell phones are not plugged and charging.

These vampire loads are the fastest growing segment of the residential electrical use market. There are many "smart" devices on the market now to address these issues and help you keep your electric costs in check, some of which will be discussed in Chapter 6. For now, it's important to make and execute a plan to structure the circuits so that we have the ability to manage critical and non-essential loads for energy efficiency. This should be part of the defined scope of work for your electrical contractor, including documentation of schematics of installation and controls.

Currently, advances are being made in the development and implementation of DC-based home energy components.[16] This technology would allow transmission of direct current from electricity-generation stations over utility grid lines with much lower line losses than alternating current (AC) distribution. Additionally, onsite renewable sources could feed current directly into the house (without requiring an inverter to convert the power to AC) to power plug-in hybrid vehicles and major appliances (for HVAC, water heating, refrigeration, laundry and cooking), as well as lighting and electronics, which would again eliminate conversion losses. We should be thinking about how we wire our homes now to allow that retrofit in the future as the technology becomes available.

Another consideration for electrical design addresses the health concerns associated with electromagnetic fields (EMFs). Exposure to electrical fields may cause health problems and will be discussed in more detail in Chapter 7. For now it is important to note that your home's electrical design should place panel boxes, laundry rooms, HVAC equipment, electronic devices, low-voltage components, electric appliances and water heaters away from main living and sleeping areas. Additionally, locate light fixtures, lighting controls and ground-fault circuit interrupter (GFCI) outlets away from areas where you frequently sit or sleep. You might even consider installing an EMF timer control device that shuts off circuits to sleeping areas on a preset schedule.

Lighting Design

Residential lighting design, which is part of the home's electrical design, should be scrutinized all on its own. The National Electrical Code sets a standard of three watts per square foot for residential construction for determining the load service required. However, some new locally adopted energy efficiency codes limit the lighting load to less than one watt per square foot. This might work for commercial applications, but remember that for the most part commercial applications occur during normal working daytime hours, as opposed to most residential lighting that is used mostly during the evening hours.

We tend to think that the best approach in residential construction is to consider the variety of usage applications in creating the lighting design; the fixtures selected for each of those applications should be based on their ability to deliver each need. Compare this to how we think about cars in terms of miles per gallon. The way to think about lighting efficiency is in lumens per watt. Recommended levels for good illumination for various types of tasks[17] are defined by lumens per square foot that can vary from 7.5 for navigation through transient areas, to 15 for general illumination in main living areas, to 75 for work surfaces. (These values vary according to the type of task and the age of the person performing it.) High-performance lighting fixtures and bulbs should supply a minimum of 40 lumens per watt, with the highest efficiency lamps rated at 60 lumens per watt or higher.

The best lighting efficiency is determined by the most effective combination of high lumens per watt fixtures and a design that provides ambient and task lighting to serve the needs of each area. As our spaces get smaller and become more multi-functional, the most efficient use of lighting is to have light sources specifically for each task and to provide controls that help us to smartly manage what is turned on at any point in time. Motion and occupancy sensors, door jamb switches and dimmers all support this.

There are several considerations you should review in your planning for lighting systems. Start by recognizing your various lighting needs and patterns of use. This requires you not only to think about how you use each room, but if there are multiple functions within each room that require separate lighting and controls. You can then determine which type of lighting will both meet each need and conserve power. Your plan should include counting on your daylighting windows to provide your primary daytime lighting source for general illumination. Of course, you'll need to plan some electrical fixtures for backup, for those cloudy days when you might need more light than windows can provide. Next, make sure that you will have overhead lighting in rooms that will require general illumination at night, such as kitchens, dining rooms and laundry rooms. Living areas and hallways may also want this option, so plan on wiring these on a separate switch. Next, plan to install task lighting in areas where you need direct lighting on a work surface, such as kitchen countertops and other work areas. Finally, think about which type of lighting control will result in the highest energy efficiency for each type of lighting.

For example, a good kitchen design has recessed can lights for general illumination when preparing meals. During the day, can you function almost exclusively with the daylighting available in the kitchen for getting a glass of water, putting dishes in the sink or dishwasher or making a sandwich on the island countertop? In the evening, cooking dinner, use the cook top vent light, the pendant light over the work surface, under-cabinet lights only if using countertop appliances, the light over the sink only when doing the dishes after dinner and overhead recessed cans only in the late fall through early summer when daylighting fades away before you get home from the office. After the cooking is finished, everything goes off ex-

cept the island pendants (you can eat at the kitchen island bar) and
then the sink light for dishwashing. After doing the dishes, everything gets turned off except maybe the light over the sink in case
you need to get a drink or to drop something in the recycle bin. You
might think that list contains a lot of electrical lighting sources and,
therefore, could not be considered energy conserving. However, we
will argue that it is how light fixtures are used that offers the best
conservation that we could hope for due to the diverse activities in
this area of the home. The key is breaking the lighting control down
to have the ability to turn lights on and off as they are needed for
specific events. Lighting controls should manage different lighting
types independently, e.g., vanity lighting in the bathroom for brushing your teeth or shaving should be separate from shower lighting.

Another consideration when making your lighting plan is the
color of the light you want and its impact on the feel of the space.
Lights come in a wide range of color outputs varying from cool
white to warm white. The cool white lamps are often also called
daylight lamps. They are best used in areas where good detail vision
is necessary, like in home offices or reading areas. The warm white
lights tend to make the space feel cozier, and they put people at ease,
so they are often used in living rooms and bedrooms. You should go
to a store with a good lighting display and take a look for yourself
and decide what you like best. The differences are quite dramatic.

More Design Considerations

Energy Modeling: Since one of your priorities is the optimal overall
energy performance of your home, you should consider hiring an
energy modeling consultant to model every aspect of your design
and building specifications. There are several recognized energy
modeling software programs on the market, including the Florida
Solar Energy Center's Energy Gauge and Architectural Energy Corporation's REM/Rate modeling software. This is the same modeling
software used to determine a home's home energy rating system
(HERS) index for energy efficient mortgages (EEMs) and code
compliance. The modeling should be performed by a professionally

licensed technician with significant experience with the software who is also familiar with different building systems, materials and design applications that can impact overall home performance.

If you truly want your home to be energy efficient, you will want to analyze how different design strategies and features, different building assemblies and different types of systems impact its performance. This modeling should begin with the design, using basic building specifications to review the impact (improve efficiency) of various component configurations.

Once we have optimized the design, this model should be analyzed for different material assemblies and then finally again for different types of mechanical systems. Do not take a salesperson's word or a manufacturer's product marketing for performance. Model the performance and know what it delivers before you finalize your design plans and specifications, and especially before you construct the building. This provides you with assurance that the design, mechanical systems and equipment that you are installing will give you the best return for your investment. It's unfortunate, but we have seen buildings fail to meet the owner's expectations on energy efficiency because of misrepresentations of system performance. Say your goal is to have a net zero energy-capable home—you want to find out whether your home can meet it in the pre-construction modeling stage, not after you have moved in and started receiving utility bills.

Building Modeling: Many architects and building designers are now using computer programs, known as building information modeling (BIM) software, as a primary design tool. These programs support the development and testing of design scenarios, helping the designer flesh out different approaches and determine how each impacts the construction budget and supports project goals. Patrick MacLeamy, chief executive officer of HOK, a global design, engineering and planning firm, has taken BIM from a design tool to a valuable resource that can be used by the design-build team in the construction phase to reduce construction costs as much as

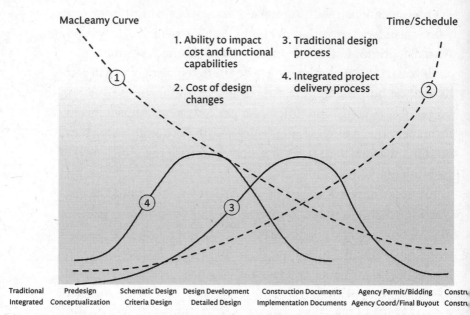

MacLeamy Curve   Time/Schedule

1. Ability to impact   3. Traditional design
   cost and functional   process
   capabilities
                            4. Integrated project
2. Cost of design   delivery process
   changes

| Traditional | Predesign | Schematic Design | Design Development | Construction Documents | Agency Permit/Bidding | Constru |
| Integrated | Conceptualization | Criteria Design | Detailed Design | Implementation Documents | Agency Coord/Final Buyout | Constr |

FIGURE 3.16. MacLeamy curve. Credit: Patrick MacLeamy, FAIA.

30 percent over conventional methods. Since typical construction costs are 20 times the cost of the building design, using the tool to develop a building assembly model (BAM) improves scheduling and subcontractor coordination and helps control costs and manage construction value. He further promotes the software's use by building owners to manage operational costs over the life of the building. Since the typical operational costs of a building are 60 times that of its design cost, he believes the model can be used to manage energy consumption and scheduled maintenance, which he terms the building operation optimization model (BOOM). The opportunities for using the model for operational cost savings can contribute substantially to offsetting the initial costs of both design and construction. BIM-BAM-BOOM![18]

Designing for Disaster: In Chapter 1, we said that considerations for site selection should include assessing the risks from natural disasters associated with specific locations. That chapter also mentioned concerns about soil stability and pest damage over time. These concerns are especially important in areas with unstable subsoils, like clay, or areas with volcanic or seismic activity that could cause events like earthquakes or tsunamis. Everyone remembers the recent wildfires that have plagued many areas of Texas, California, Colorado and many other western states, the 2011 earthquake that rocked Virginia and further up the East Coast, and the record storms and hurricanes that have taken countless lives and caused billions of dollars worth of damage across the southern and northeast corridors of the US. Numerous types of natural disasters should be evaluated in your risk assessment, including those we have already mentioned and others such as avalanches, blizzards, landslides, floods, tornadoes, extreme heat and drought .

Building science has advanced our ability to deal with these various risks, through incorporating both design strategies and improved building materials. Hazard mitigation methods can reduce the incidence of loss of life, property and function due to natural disasters. The National Institute of Building Science Whole Building Design Guide website (wbdg.org/design/resist_hazards.php) provides up-to-date references on resources to address these concerns. The frequency and severity of recent extreme weather events has heightened the need to design structures to withstand them. These methods can save tremendous resources over time, as well as human lives.

Designing for Deconstruction—Will You Ever Want to Remodel?
Again, think about your home over the next hundred years. How many of you know people who live in homes that are 15, 20, 25, 30, 50 years old or older that are constantly remodeling them? What are they changing? Floor plans that no longer work well, out-of-date styles, worn out finishes. How many times does a minor renovation turn into a big remodel due to the difficulty of removing old installations without major damage to other parts of the structure?

What if we rethink how we install building components to make it easier to remove them in the future? What if we could provide the materials for future generations to build with, eliminating the need for them to harvest virgin resources? What if we could design for deconstruction?[19]

You could go so far as to build interior, non-load-bearing walls after the building ceilings are finished and painted, so they act more like movable partitions (as we do in commercial spaces) if you want to reconfigure rooms in the future. Even if you don't go that far, consider designing your kitchen and bathroom cabinets runs to be open-ended. This means that at least one end of the cabinet run is not enclosed by a wall. This will save you money on your initial cabinetry costs, because finished dimensions are not as critical as they would need to be to fit within two finished walls. It also saves a trip charge for the cabinet vendor's service tech to drive out in order to get those exact dimensions after the drywall has been hung. It also allows for more flexibility so when the day comes that someone removes the cabinets in a remodel, they are more easily reusable in other spaces.

There are many opportunities to design for future deconstruction, salvage and reuse. Try to look at buildings from the end-of-life perspective. In the future, as resources become scarce and in higher demand, having salvageable materials will significantly increase your home's value.

Added Value — The Integrated Project Team Approach: Although you can use green building materials, systems and finishes on a packaged home plan, in essence you are missing the boat if you don't engage the services of a local reputable *green* architect or experienced design/build firm. Be sure to make it clear that your goal is to design in such a way that you reduce the amount of building materials needed for each system in the home. This includes designing the frame, roof, foundation and mechanical systems so that you deliver the most performance with the least amount of materials. Share the exact strategies outlined in this book that you expect to be used on your project, emphasizing their importance to achieving your budget goals.

When you think that there is nothing more that you and your architect can do to improve the design, it's time to turn it over to the project team for their review and input. At this point, you should have already interviewed and hired a reputable green homebuilder to oversee the project, and he/she should assemble their framer, mechanical contractor and plumber for a design team meeting. Each contractor should be provided with a scope of work specific to their trade. It outlines their responsibilities and details any specific methods that are to be used on the project. This will give them an idea of how to best integrate their work with other aspects of the project.

Building commissioning should start during the design and planning phase of your project. One of the functions of a building commissioning agent is to assure that the green goals of the project are met at each milestone of the process. This means that preliminary rough building plans are reviewed for how well they support overall goals, as are the methods and materials proposed by the project team. The use of continuous and contiguous air, moisture and thermal barriers (discussed in Chapter 4) should be something that is documented in the building specifications and then verified on design plans by the agent. Considerations for variations in wind and pressure differences for each exterior wall and roof assembly should also be included on the plans. The agent should be tasked with the responsibility of inspecting for all aspects of proper installation of these various control layers during construction to confirm that they are installed as designed.

It is important that you communicate your goals and these strategies to your builder and the various members of your project team. Make certain they are documented in the construction bid packages for various trades, including specific details that will result in fewer materials needed and wasted (see Chapter 5 on construction waste). Fewer materials also results in less labor for installation. This results in lower base construction bids and leaves more money in the base budget for high-performance upgrades. If you follow the methods recommended in this book, you should see lower base bids from the foundation contractor, the framer, the plumber, the HVAC contractor, as well as from the various interior finish contractors. If you are fortunate enough to have experienced contractors who are familiar with green strategies on your team, it is good to get their advice and guidance on the latest technologies and products that you should consider, but be sure to price these out separately so that you can analyze the potential return on each investment. You should also discuss the types of building envelope systems under consideration and find out the pros and cons of each. The discussion should encourage contractor input on options, cost estimates or cost savings, including how each impacts other component selections and costs. For example, if you are interested in spray foam insulation at the roof rafters in order to bring the mechanical equipment and duct system within the thermal envelope, then the mechanical contractor should acknowledge that this will likely reduce the loads on the equipment, as it effectively lowers the required British thermal units (BTUs) of equipment capacity to heat and cool the home, thus reducing the size and cost of equipment and duct runs.

Or, seeing that you are building in a location that does not have gas service available, he might suggest an electric heat pump unit in lieu of a sealed combustion propane furnace. The mechanical contractor might charge a little more for the heat pump unit, but the builder will recognize that the savings of not having to run gas lines or a flue through the roof will offset those costs. And you should recognize the operational cost savings of electricity versus propane, especially if you have the opportunity to install solar PV.

It's also a good idea to have the contractors review the design to recommend changes to improve system performance. It's important to listen to their ideas, concerns and suggestions before finalizing any design plan. The mechanical contractor might point out that the design does not leave a good path for ducting air supplies to various rooms for the air conditioning and heating system. Together with the framer and the truss company, they may recommend placement of additional chases, or changes in ceiling height or the direction that trusses are run. The plumber should provide input on the centralized location of the water heater or installing an on-demand pump to reduce water waste and improve convenience to remote wet locations. This is the integrated project team approach to design. It can result in significant improvements in energy and water efficiency, building performance and reducing costs.

Of course, other contractors might participate in design plan review—the more the merrier. The structural engineer, foundation contractor and roofer might suggest cost-saving strategies, and this alone can be worth paying the architect to revise drawings. We're always amazed at the number of homeowners that do not realize it wasn't the green features that added cost to their house, it was the extreme inefficiency of their design. Inefficient design can add costs in materials and labor across the board, from foundations to framing to plumbing to roofing to finish out. The greatest cost savings you can achieve is through good design, and those savings can easily pay for upgraded equipment, amenities and finish out.

Payback on Design Elements

Your best opportunity to reduce the construction and operating costs of your home is through design strategies that provide for passive solar, natural ventilation and water management. Passive solar features alone can reduce heating and cooling loads 60–80 percent, reducing your utility bill 40–80 percent.[20] These savings will come month after month, year after year, so long as the building is standing, and do not depend on the installation or operation of expensive mechanical systems that might break down over time.

Some design features are more difficult to assess and/or have longer returns on investment. Some will contribute to improving comfort, which results in energy savings, so you can actually see operational dollar savings through energy modeling. By taking advantage of and working with natural features, you will be able to reduce the size of mechanical systems, which directly impacts your construction costs.

You will also realize immediate savings on your construction bids from a simpler design, compact plumbing and HVAC design, as well as standardized structural elements. And engaging your project team's input on design and system integration can lead to incredible cost savings, both initially and over time. If you fail to do this, you may actually spend more money trying to overcome poor planning. Design strategies also increase the home's durability and value over time by minimizing damage in the event of a natural disaster and incorporating end-of-life salvage values. Efficiencies in design can also reduce water waste and increase conservation, as well as manage electrical loads for energy savings. And if that were not enough reason to focus first and above all else on these strategies, they are also the least expensive path to the net zero benefits discussed in Part Two of this book.

CHAPTER 4

Building Products and Materials: Shades of Green

Our choices of building products and how well we use them contribute greatly to achieving our goals of energy, water and resource efficiency, and healthier, more durable and affordable homes. But sometimes those choices also have impacts on a much larger scale. In this chapter, we will look at building materials in terms of "shades of green," recognizing how well they contribute to a green home, but also measuring their benefits in terms of the larger picture of the world we live in and how our purchasing decisions affect it.

Global Citizens

We live in a global economy; many of the raw materials and products we use in building are sourced from various regions around the world. We've all heard the stories of sweatshops in impoverished cities and mining and manufacturing operations that pollute air and water. Methods of exploitation of natural resources have ravaged natural environments around the globe, displacing and, in some cases, endangering the indigenous people and wildlife that once occupied those areas. Certain ecosystems, like oceans, soils, rainforests, wetlands and native grasslands sequester vast amounts of carbon, keeping them from being released into the Earth's atmosphere. Human activities related to the disturbance and destruction of these environments have played a major role in man's contribution to global warming. When we purchase a building material, do we ask where it came from and what the impacts were to get it to our market?

It is evident that many people must believe that Big Brother is overseeing our safety or at least monitoring how the companies that supply our goods do business. With the exception of a few regulatory and licensing agencies, this is just not the case. Many building materials come from areas of the globe that have no regulation of mining and forestry practices. Growing world population numbers indicate a continued trend toward depleting natural resources at alarming rates. The planet itself is suffering from ozone damage and toxic waste dumps, threatening numerous ecosystems to the point of collapse.

In this chapter, we will raise some important considerations to help you make better purchasing decisions. Green products promote a conscious effort not only to increase building efficiencies, but also to reduce the impacts on the larger community, whether locally or globally. To help you become educated consumers, we want you to know the right questions to ask to get...

The Whole Truth

The market is overflowing with new technologies and new products, constantly expanding with more choices. The construction industry is also in transition, and, we are happy to report, it reflects green going mainstream. And it's not just the building industry, it is all of the supporting industries as well. Home inspectors are learning to inspect for green features in homes, real estate professionals are learning to market them, and product manufactures are improving building materials to build them with. But it can take time to turn big ships around. Being green is popular now. And everyone wants to be popular, don't they?

"Greenwashing" is the term coined to describe false claims made about products' green features. Many manufacturers have jumped on the green bandwagon by looking at every possible attribute that might make a product fit into some green claim and then carefully wording their claims to trick you into thinking their product is green. Research studies have shown that as much as 98 percent of all green marketing claims are either false or misleading.[1]

Greenwashing can take many forms. We have seen product manufacturers referencing credits available from green building programs in their product advertisements without ever saying that their product meets any of those criteria. We have seen creative little green logos on labels intended to give the impression that some credible authority has verified their claims. Although some attempts are being made to police marketing claims being made about materials and products, there really are not enough resources available to administer this function at every level of the marketplace.

Remember, all marketing is intended to sell you something. We all have to do our homework. Even those of us who work in green building have to rely on credible third-party sources to verify, test and compare products. Most third-party verification sources specialize in particular product groups or green features. For example, Green Seal provides testing, reporting and certification of products that have a low impact on indoor air quality. Other credible independent entities review and certify products for energy efficiency, water conservation, recycled and toxic ingredient content. So, you must be certain that any product labeling claims are from a reputable third-party source.

Building Product Research

There are complete books on the market that give detailed descriptions of the vast array of building materials available. As soon as a book is published, though, it is out of date, not only due to the continued invention of new products and innovations in existing ones, but also due to the results of continued research into just how green all of these products are. Additionally, not all products are available, or even appropriate for use, in every market. Your best resource for building materials is through a reputable green architect, home-builder or building supply outlet that specializes in these types of products and has a knowledgeable staff to help guide you.

There are very few perfect green products. Generally products are categorized by certain attributes that are considered green, and some products will have more green attributes than others. So, it is important to recognize products by shades of green. As an example, look at one of the earliest recognized green products around—straw, as used in straw bale construction. Straw is a waste product of cereal grain production, and so, as we will discuss a little later in this chapter, that's a good green attribute. It's even better if the cereal grain is grown nearby, as local material sourcing is also a great green attribute. In using straw for your building system, it can be durable and provide good thermal performance. It also does not off-gas toxic fumes, so it will not affect indoor air quality. In fact, we would say straw is just about dark green—as perfect a green product as you can get if it meets all of these criteria. Unfortunately, we don't build many homes with it.

On the other end of the spectrum, let's look at recycled glass content fiberglass batt insulation. Yes, it is recycled, but batt insulation (discussed later in this chapter) typically does not provide good thermal performance due to installation errors, may have added urea-formaldehyde binders that will off-gas toxic volatile organic compounds (discussed further in Chapter 7), affecting indoor air quality, and probably is not produced locally. So many would consider this product pale green.

The reality is that there are many trade-offs in the designation of green products. Some aspects can make or break your expectations of the benefits your green home should provide, while others just represent added value. What you need to know is which attributes are important.

Main Considerations

You should start your analysis of building materials by thinking about what materials are appropriate for you to use on your project and specifically in your climate. Just as construction methods are climate specific, so is our use of building materials. As you will see throughout this chapter, some products are specific to hot and humid climates, while others are best used in cold, arid or marine climates. In some instances, it's not the product itself, but how it is used that is determined by climate type. Also, we need to be sure we have checked what applications are appropriate for our site conditions and building design.

Building science research continues to advance, ushering in new technologies and product developments to the market, now at exponential rates. It is important that we make sure the contractors that are installing them are properly trained so that we get a good installation quality. Sometimes it's not the product itself but how well we use it that makes it green. If you consider yourself an early adopter and like being the first to own the latest gadgets or the most expensive model with all the bells and whistles, you can bet you will spend more money than this book suggests. Just about all new technologies must recover the initial cost of research and development and their start-up manufacturing costs, so you are better off if you can wait until prices stabilize. You will also find that time saves you the grief of working out the kinks that are inherent to new product development.

We also need to think about how well products support our energy and water efficiency goals. And, as was the case with designing for risk assessment in Chapter 3, you must specify building materials able to stand up to your assessed risk for tornados, hurricanes, floods, termites and other potential hazards. Certainly wood siding would not be a good choice if you live in an area with a high termite risk.

Performance and Impact on Health: It is equally important to recognize what characteristics you are looking for in each product category, so as not to give more weight to a green attribute than the value you actually need from the product. Your review of any building material should begin with an analysis of what function you need the product to perform and any concerns you should be aware of with using it.

For example, with any insulation material, you first want good thermal performance, and second you do not want any product that will adversely impact indoor air quality and your family's health. Any products used within the wall, floor or roof structure or interior finishes can off-gas toxins into the living space, impacting the health of workers manufacturing and installing them and, of course, the home's occupants. This will be discussed in depth in Chapter 7, so for now we will only recognize that it is an important consideration for selecting some building materials. Once a building material has passed those two tests, you can consider other green attributes (recycled or waste content, locally sourced, etc.) in your final selection. We will review the important considerations for each type of building material later in the chapter.

Longer Maintenance, Repair and Replacement Cycles: We do not generally set aside money in our monthly budgets to address maintenance and repairs over time. But if we did, and we had to amortize those lifetime costs into a monthly budgetary allowance, we would see the value of investing in better-quality products to begin with. A primary goal of building science and green building is to achieve a home that is durable with less maintenance. Part of how this is achieved involves attention to the details of what it is built of as well as how the house is built.

Choose materials that will hold up over the long run and that are less susceptible to damage from pests and weather events. Natural stone, cement and clay products provide excellent durability when it comes to the exterior façade of the home (from siding to roofing materials). These products also work well in the interior (ceramic tile or concrete flooring, clay plaster walls and natural stone or other solid-surface countertops).

We, as a society, need products that last longer. It is senseless to continue to use our energy and natural resources to manufacture more, simply due to our failure to choose a product initially that was durable enough to last. We cannot continue to produce products that are cheap and disposable and just throw them away. Our planet, Earth, is all that we have, there is no "away."[2]

We are discarding a greater volume of products daily than Earth can decompose and regenerate back into new resources.[3] Choose materials that reduce the volume of resources required over time. Tile floors do not need to be replaced as often as carpet flooring. Engineered wood products experience less twisting and warping, so there is less building failure over time that requires replacement resources.

Aesthetics: Choose materials that are pleasing to look at—classic styles, not fads. Many resources are wasted in remodeling or redecorating because poor initial choices were made. This is an especially concern for interior finish products, many of which off-gas toxic fumes for years after installation. Just about the time the initial product ends its toxic period, it is out of style and replaced by a new product, which has to begin the off-gassing cycle all over again.

Life Cycle Assessment: One of the verification tools we use to determine a product's shade of green is called a life cycle assessment or LCA for short. As the term indicates, this type of assessment provides an analysis of something over its entire lifetime. For building materials, this means analyzing the impact of extracting or harvesting raw materials, processing or manufacturing the product, installation on the jobsite, the maintenance and operations through the product's useful life, to the end-of-life removal and whether it can be repurposed, recycled, or ends up in the landfill. The assessment also includes all the energy embodied in the product from the generation of the energy required to make all these processes happen and in the transportation required to get the product from one process to the next.

LCA studies have been done on a wide range of topics, like how much energy, water or natural resources something consumes over its lifetime, or maybe the impact of living in the home on our

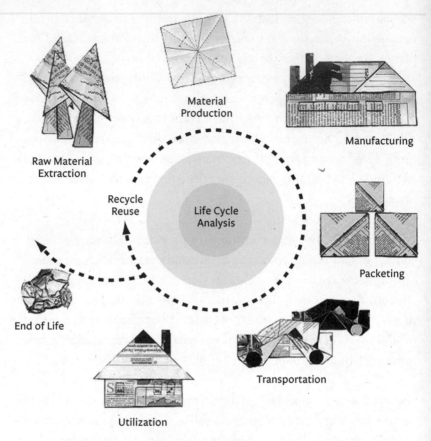

FIGURE 4.1. Life cycle of assessment of building materials.

family's health over its lifetime or the total cost of ownership over its lifetime. Most of you will never have the opportunity to see any of these studies in your research of building materials. What we would like for you to take away from this discussion is that you can pretty much do the same type of analysis yourself, just by giving some thought to how the products are going to perform over time, how they benefit your long-term goals, where the raw materials come from and what processes are required to get them to you in a usable state and, finally, how those processes might affect the world that we live in over the long run and the generations to come after us. This requires consideration of a product's durability, life expectancy and ease of maintenance. It also requires us to think about making more sustainable choices, such as those in the following discussions.

Man-made vs. Natural Products: Products made from natural resources deplete those resource reserves. Just as fossil fuels are in limited supply on this planet, so are other naturally occurring materials such as lime, which is used to make Portland cement, wood used for everything from structures to flooring, and gypsum, used to make drywall. In the case of natural products, we also need to consider if the raw material source is renewable or finite.

Sustainably sourced wood lumber for framing is grown on farms, and many of the companies in this industry have already realized that to stay in business they must replant after every harvest to assure future supplies. However, many exotic wood species come from regions of the world where these practices have not been adopted. For example, ipe is a wood species that is only found in Central and South America. Most of it on the market today is being illegally harvested in Brazil. On average, only one or two ipe trees grow per acre of rain forest,[4] yet the entire acre is cut to harvest these two trees. When the rain forest is gone, it's gone forever.

On the other hand, man-made products generally have higher embodied energies due to the manufacturing and processing. This includes products made from synthetic materials, such as plastics. It also includes products made with recycled content, so there is a trade-off between the energy required to manufacture new synthetics and the energy needed to collect and ship recycled materials back to the manufacturer and reprocess them into a new product. We need to create economically profitable paths for manufacturers to recover these waste streams.

There are many green man-made products on the market, too many, in fact, to even attempt a list. It is important to recognize that some products are both natural and man-made. This refers to materials that use ingredients that are natural but must be processed to be usable, such as products made from glass (silica), metals and gypsum (drywall). These products both deplete natural resources and have higher embodied energies. Recycling these products protects against excessive natural resource depletion.

Resource Efficiency

With regard to resource efficiency, we need products that consume fewer high-value natural resources both initially in their manufacture and over the long term. So, we really like products made from rapidly renewable resources, salvage, recycled or waste materials, or that have been otherwise repurposed from existing, not virgin, resources. Or products that serve more than one purpose, like concrete foundations as finished floors, so additional flooring products are not required. And products that are reusable or recyclable at the end of their lives, saving further depletion of our natural resources in the future.

Salvaged Materials: Materials that have been reused from deconstruction or demolition projects can find new life cycles as either the same application as before or in a completely different application. Wood flooring is the most commonly salvaged and reused building material. Another is brick masonry, which is often reused as hard surface paver materials (e.g., for driveways and walkways).

We have also seen many affordable projects constructed or remodeled with numerous salvaged materials, including cabinetry, doors, bathtubs and sinks and even scrap lumber used as paneling. Sources include resale shops, like those run by Habitat for Humanity to raise funds for their own construction of affordable housing units, as well as similar facilities run on inventories created by demolition and deconstruction companies. There are even extreme homes constructed entirely with salvaged materials, such as Earthships, which use old tires filled with dirt and plastered into retaining walls using adobe or cement to form the building shell. In some areas of the country, lacking any facilities within a reasonable distance that might be able to recycle these tires, they could have easily ended up in a landfill. We have also read of and seen photos of structures built entirely from glass bottles and mortar. Are these homes green? From a resource efficiency standpoint, yes.

If you think about it, before mankind had the ability to ship various materials across the country or around the world, all homes were built out of local available materials, whether they were virgin resources or reused existing components. In fact, in those days, there were no such thing as landfills and new resources were scarce, so things were used as long as they were useful. Are we resurrecting this into a new trend?

Materials That Have Recycled Content or Waste Content: Sometimes it is difficult to estimate all the energy required to salvage, transport and repurpose materials into new products as compared to that used to harvest raw resources in order to determine which is greener. We have already discussed the impact on the environment from many of the practices used to extract virgin resources. We also know that when natural resources become depleted, their cost increases, affecting the affordability of the material and thereby impacting the overall cost of housing. So, in the end, recycling wins.

Basically, there are two types of recycled content for manufactured goods: post-industrial (waste products from the manufacture of another product) or post-consumer (waste product of a product already used). An example of post-industrial waste would be oriented-strand board (OSB), made from wood chip waste created in the milling process of other wood products. An example of post-consumer waste would be those flimsy plastic grocery bags that can be recycled into carpet and composite decking materials for patios. Post-consumer waste is the target, as most of that ends up in landfills, so recycling it not only saves virgin resources but also reduces the landfill mass that contributes to greenhouse gas emissions.

There are many other products on the market that contain recycled waste materials, everything from recycled glass or paper countertops to recycled rubber tires made into a mulch product. Building products to look for with recycled content include metal roofing, roofing asphalt shingles, sheathing and house wraps, window frames, roofing materials, composite decking and ceramic flooring. Insulation products are available that are made from recycled materials (cellulose is recycled newspaper, cotton is from recycled blue jeans, and, of course, recycled fiberglass). As was mentioned earlier, many products can be recycled back into production of the same product again.

Rapidly Renewable Materials: Sustainably sourced fast-growing species of bamboo and cork are popular choices for many durable building materials. They should be considered as alternatives to old-growth, more vulnerable wood products. Growing them does not require chemical fertilizers and pesticide treatments nor do the products made from them need periodic maintenance, such as painting, staining or sealing. These materials are commonly used to make cabinetry, flooring, countertops, insulation and trim. Look for products from certified sources that use formaldehyde-free binders (see Chapter 7).

Bio-based Materials: Bio-based products are generally from rapidly renewable resources, and many represent waste streams from other industries (food or fiber crops). These materials have very low embodied energy, since the energy to grow them is attributed to the primary crop. Therefore, it is only the processing and transportation energies that add measurable value to any product made from them. From corn starch to soy beans, wheat-straw shafts to sunflower stalks and rice hulls, these materials are finding their way into many building products. Look for bio-based solutions in wallboards, drywall, ceiling tiles, foam insulation, cabinetry, adhesives and caulks, as well as those used as binders in the production of other building materials to replace toxic chemicals.

Indigenous Materials: Raw materials sourced or products manufactured within 500 miles of the project site are considered green because they have lower embodied energies from transportation. Look for materials harvested or mined close to your area, includ-

ing wood products and aggregates, and check for local manufacturers in your region. It also generally uses less energy to ship raw
materials some distance to a local factory than to ship finished
goods over great distances. Many national manufacturers have
production plants in various locations across the country and
would be happy to ship from the closest location to your project if
requested.

Products That Require Minimal Finishing: More and more manufacturers are developing a competitive edge through product improvements that offer more durability and less maintenance. Fiber-
cement siding materials are now offered with integral color[5] finishes
providing up to 50-year warranties on products that are guaranteed
not to need painting for 15 years. The same is true for American clay
plaster materials, available in almost as many color palettes as interior paint products. For your patio, composite decking materials
come prefinished, never requiring resealing or restaining, unlike
their wood counterparts. Remember to think about how often any
finish material (exterior or interior) will need maintenance or repair. Choosing products that have less frequent cycles saves money
and resources over the long run.

Selecting Building Materials

It would be impossible for us to cover all of the building materials
on the market in this book. Now that we have described the main
considerations for selecting building materials, it is necessary to put
that into context for each type of material. We are going to look at
the major building component categories and attempt to provide
that guidance for each, starting with the outside walls, foundation
and roof and work our way in.

Remember to focus on the primary function or performance
that you need from the product, as well as any concerns from using
it. All materials used within the walls of your home and as interior
finishes can have health impacts, while others may have environmental impacts. In our discussion of materials in this section, we

will just acknowledge in which product categories these considerations are important.

Foundations and Hardscapes

We want strong foundations, and they should be designed for your site, so this means boring soil samples and engineering accordingly. Which type of foundation is right for you will depend upon your site and soils. We tend to prefer pier and beam or basement foundations as they are much more material efficient than a slab foundation and provide the ability to access plumbing as needed over time for repairs or renovations. If radon, a deadly radioactive cancer-causing soil gas, is not present in your area, you should consider installing a moisture barrier over the soil and sealing up the crawl-space and installing insulation at the foundation walls. If radon is present, you should ventilate that assembly with an exhaust fan to create a negative pressure between the crawlspace and the house. If you go the slab route in a risk area for radon, you'll want to install a radon ventilation system as part of your foundation design. There will be more on this in Chapter 7.

Also, if you go the slab route, consider stepping the foundation for any major changes in grade, as this also will reduce the amount of materials required. We've seen foundations drop 17 feet off the side of a hill, taking a lot of concrete and fill to assemble. If you must install piers or footings anyway, better to use beams for floor support and then enclose the parameter with ICF panels (insulated concrete forms). This assembly creates a great storage facility that can be used as a mechanical room.

Make certain that your foundation contractor reuses old form boards from one project to the next. There is no reason to see this lumber go into your trash bin. Also, your contractor should be using non-petroleum-based release agents on those form boards, as the substance that keeps the concrete from sticking to the wood. This alternative product will not leach toxins into your soils as it is washed off. Also, remember that the concrete sticks to the inside of the truck, so when the truck driver washes out the tumbler, this sends toxic chemical binders into our soil and eventually water systems. There are concrete washout systems that can be used to treat the wastewater and reclaim any aggregates from the waste.

When you can, limit impervious surfaces, including driveways, walkways and patios (hardscapes). These areas reduce the land available for precious rainwater to infiltrate onsite, providing natural irrigation. Consider permeable alternatives, such as permeable concrete or pavers. With these materials, you must make sure the contractor prepares an aggregate bed, per manufacturer's guidelines, and protects the area from future contamination that can be caused when erosion carries soils, compost and mulch materials over the surface, as these can clog up the drainage.

Finally, choose light-colored materials for all outdoor hardscape surfaces to reduce the urban heat island effect. If this is not possible, plant fast-growing shade tree species that will provide shading to these hardscapes within five years. Shade arbors and pergolas also can be used as shading devices for the hardscape, as well as providing a shade buffer zone for the adjoining side of the house.

Structure

We want our structures to be durable, low maintenance and have high thermal value. The framing materials that you select should provide a good thermal envelope and be resistant to pests and weather damage. Remember to refer to your risk assessment that was discussed in Chapter 1 to determine which materials and systems will address the risks likely for your site over the life of your home. Special materials may be required if you are building in areas with concern for hurricane, tornado, or earthquake activity.

There are numerous choices that you can utilize to achieve a darker shade of green on your project, including improved framing techniques, engineered products, as well as alternative and natural building systems. The latter two are both resistant to weather and pest, as well as providing exceptional resistance to fire, mold and seismic risks. These are durable, long-life expectancy systems with high thermal and acoustic ratings and will be discussed further after we look at traditional framed structures.

Sustainable Framing Materials: The majority of all homes built today are wood framed. When selecting wood products, it is important to rely on credible third-party certifications for lumber sources. The two main certification entities are the Forest Stewardship Council (FSC-stamped wood products) and the Sustainable Forestry Initiative (SFI-labeled products). FSC-stamped wood products are harvested from forests managed according to the guidelines of the Forest Stewardship Council, a non-profit organization. This means that trees are harvested under strict regulation, respecting the rights of indigenous people and laws of forest lands, that fair trade practices protect local economies and that seedlings are replanted to replace what's harvested. These products require a chain of custody, meaning that every entity that handles the product from the time it is harvested until it is delivered to your jobsite is certified to assure that program requirements are met. Compliance is verified by audit visits. SFI products do not carry a chain of custody; while compliance with their guidelines is voluntary and not subject to the same verification audits, the intent is similar.

Engineered Products: When it comes to structural members, engineered products include glue-lam beams and headers, as well as roof and floor trusses. They are engineered for strength, so they are a great alternative to solid sawn lumber that requires older-growth trees. Other engineered products include finger-jointed studs and trim and oriented-strand board (OSB), and medium density fiberboard (MDF).

The engineering design of all of these products provides significant improvements in structural stability, as the products are straighter and have fewer defects. Since knotting and grain abnormalities have been removed, you don't get the twisting and warping problems associated with solid-sawn lumber that resulted in structural and cosmetic problems over time that take more resources to repair. This means that even though some of these materials may carry a little cost premium, they result in having to use less structural framing than when solid sawn lumber is used exclusively, resulting in no net cost premium.[6] In fact, using engineered trusses and wall panels (see Chapter 5) can reduce the amount of lumber required by 25–35 percent and the time spent by 30–50 percent, resulting in a net cost savings of 16 percent over conventional framing.[7]

Metal Framing: As an alternative to wood framing, metal offers long-term durability and resistance to weather and pests. Metal is energy-intensive to produce, but it is infinitely recyclable. However, since thermal performance is one of our primary objectives, it is worth noting that metal is a great conductor of heat. Since framing makes up about 25 percent of the exterior of your home,[8] steel will conduct about four hundred times as much heat into or out of your home as wood framing does. This can reduce the average R-value of the wall as a whole by over one half![9] To reduce thermal bridging when using metal framing members, it is important, and required by the building codes, to use an insulating sheathing continuously around the exterior of the home. (Note that wood framing is also subject to thermal bridging, so using a rigid foam sheathing on top of your structural sheathing is always a good idea). The thickness of the insulating sheathing foam is dictated by the severity of your climate (and is specified in the building or energy code used in your region) with ranges from half an inch to two inches, with a minimum of R-4.2 required even in the most hospitable of climates.

Natural Building Systems: Natural building systems have been around since life began on Earth. Just as birds build nests from sticks and feathers, man has constructed shelters from combinations of earth and vegetation since we migrated out of caves. These types of structure have evolved throughout history based on what materials were available locally, the skill sets of indigenous people and, of course, trial and error to address various previous building failures or shortcomings.

Today, civilized society tends to prefer man-made materials, mostly because we have become a society of specialists. We no longer depend on our neighbors to help us raise the barn, we just hire an architect and builder and write checks to cover construction expenses. And, in doing so, we tend to try to go with the easiest and least expensive building systems. Craftsmanship has given way to volume production efficiencies. But there are still natural homes being built, and again, these types of construction tend to use the most perfect green products, with low embodied energy and impact on indoor air quality (IAQ).

There are a wide variety of natural building systems in use in construction today, including straw bale, rammed earth, cob, cordwood and adobe, each with recognizable benefits. It is up to you to research their suitability for your climate, as well as verify actual thermal performance and other benefits. There are entire books on each of these building systems, so we recommend that you do your own research into any specific type that you are considering. Energy modeling can assist you with this, but you should seek out qualified local referrals from those who have used and are experienced with each type of building system you are considering. Our goal is just to make certain you are aware of them and to encourage you to consider these alternative choices to wood-framed construction.

No book on green building would be complete without recognizing the passive strategy offered by earth-sheltered construction. Raw materials from the earth are the most abundant building materials available, and, if locally sourced, should have the best life cycle assessment. Using the earth's ability to act as thermal mass[10] helps to mitigate heat transfer in climates with high diurnal swings. There are, of course, other considerations that must be addressed, primarily fresh air ventilation, daylighting and moisture control. Whether you are considering above-ground earth-bermed construction, below grade (or built into a hillside), or living in a cave, each has its own structural challenges and requires expert guidance to assure safety and durability.

Alternative Building Systems: Alternative building materials markets are dominated by structural insulated panels (SIPs), autoclaved aerated concrete block (AACs), and insulated concrete forms (ICFs). These materials are generally considered durable, offer above average thermal performance, and generate little construction waste. Depending on the materials used to manufacture them, they may or may not contribute to indoor air quality issues. They are resistant to fire, mold and mildew, pest and weather damage, and offer good acoustic ratings. Some alternative systems, like structural insulated panels (SIPs) can be made from agriboard (straw board) and bio-based foams that have less effect on air quality. AACs also have low impact on air quality. Again, do your research beyond what the manufacturer is promoting about their performance and find a builder experienced in these materials to advise you on what is appropriate for your area.

Pest Control for the Structure: If you are located in an area with a moderate to high risk for termites, you will want to look at an integrated pest management (IPM)[11] plan to reduce the risk of damage to your building over its lifetime. Building codes may only require that the sole plate (the bottom framing member that touches the foundation) be treated for termite resistance, but that is not enough to stop an invading swarm of termites. For a few dollars, you can go to your local pest product supply house or go online and pick up a gallon of borate spray and a two-gallon pump sprayer and spray all of the wood framing at least three feet up from the foundation.

Most termites are subterranean, meaning they will crawl under your foundation and come up into interior walls through plumbing penetrations in your floor. To stop this, there are metal mesh systems on the market that can be installed around penetrations to cut off this point of entry. Your mason should install a fine stainless steel mesh or steel wool in your weep holes on masonry walls to negate this point of entry into wall assemblies for any pests. Some very fine sand types (16 grit) can also be used around plumbing drains and such to prevent termite entry. Other species of termites can fly, so if you have a vented attic system, make certain all exterior vents are covered by a screen to prevent entry.

Insulation: What is most important is thermal value, with impact on indoor air quality second. For superior thermal performance, installation quality is as important as the product that you choose. Insulation is made of materials that are bound together with thousands of tiny air gaps in the material itself, and these air spaces are what mitigate heat transfer. If the material is compressed, those little air pockets close up, or if it does not totally fill the cavity it is installed in, leaving gaps and voids, it quickly loses its ability to perform as intended.

To minimize the impact on indoor air quality, choose insulation made from materials that minimize off-gassing of volatile organic compounds (VOCs). Although the way fiberglass insulation is manufactured creates some urea-formaldehyde in the products, many fiberglass insulations contain added formaldehyde in the form of a binder to hold the fibers together. You should avoid these products with added urea-formaldehyde, as this is a toxin that can off-gas for years.

Properly installed spray foam products may off–gas most of their toxic fumes in the first 48 hours or within two weeks of installation,[12] but a few people with multiple chemical sensitivities may still notice adverse effects for much longer periods. For these people, great care must be taken in the selection of every material used in the construction of the home and expert guidance is advised. Once you've addressed the thermal performance and lower impact on IAQ, then if you've found a product made with recycled content, consider that a bonus.

To determine thermal performance, the ENERGY STAR for homes program grades insulation installation quality in terms of

Grade I, II, and III. Grade III is what we typically see with batt insulation, defined as having more than ten percent overall gaps, voids and compressions. Remember, each wall cavity may have bracing, windows, doors or other framing that makes it a non-standard sized opening. Batt insulation, on the other hand, comes in standard sizes. So, they are either cut to fit (and usually not very exactly) or they are crammed into a smaller space than they were intended to fit. The former can result in gaps and the latter in compressions. These defects in installation quality reduce your effective R-value by at least 42 percent, according to studies[13] conducted at the Oak Ridge National Laboratory by Dr. Jeff Christian and Jan Kosny.

Batt insulation is also difficult to install without compressing it behind wiring and plumbing runs through the cavities, and it is not easy to cut to fit well around electrical outlet and switch boxes without compressing it. Installers are generally paid by the job so they try to do as many homes a week as they can. To make installations go faster, they will try to compress the insulation behind the obstructions and often don't take the time to cut carefully. Then there are the gaps and compressions in cavities that are so small the installer cannot get his hand in for a proper fit. If you insist on getting a good installation of batt insulation, your contractor might be able to achieve a Grade II. This is defined by ENERGY STAR as less than ten percent but more than two percent overall gaps, voids and compressions. It is almost impossible to get to Grade I with batt insulation, as this is defined as less than two percent overall gaps, voids and compressions. Less than two percent overall represents a near-perfect installation.[14]

Trust us, you are not going to get that kind of quality unless you are dealing with an extremely reputable firm, one that knows that you have hired an independent third-party inspector to review the quality of their work. Otherwise you have to specify this in your insulation specification or in that contractor's scope of work agreement. Your best bet is to go with a densely packed blown-in-blanket (BIB) system, a damp spray cellulose, a rigid foam board, a spray-in-place formaldehyde-free fiberglass or a spray-in-place foam. These systems give a total fill to each cavity, and are dense enough to resist air movement and convection loops.

Batt insulation can be made from fiberglass (silica) or natural insulation materials like recycled blue jeans or cotton fibers. BIB systems use a fabric shield, stapled to the cavity studs, filled with some type of blown-in insulation material. Better choices in these include cellulose (recycled newspaper treated with borates, which is preferred, or aluminum sulfate added for termite resistance and fireproofing), rock wool or formaldehyde-free blown-in fiberglass.

Both open-cell and closed-cell spray foam insulation products are also total fill systems. Some foam products have a small percentage of bio-based content made from natural, living materials, such as soy-based foams. Foam products act as their own air barrier, so you can expect the best overall performance from this type of insulation.

When using spray foam for sealed attics, your choice in products should depend on your climate type. Spray foam insulations are defined by the type of cell formed by the foam and by the density per cubic foot. If you live in a hot/humid, hot/dry or mixed-humid climate you can use either open-cell one-half-pound density or closed-cell two-pound density spray foam. If you live in a cold or severe cold climate, you must use only the closed-cell foams for its low water vapor permeability to avoid problems with condensation in the winter.

The difference between the two products is their permeability to water vapor. Open-cell foam is far more vapor permeable at 5–7 perms than closed-cell foams that are less than one perm. This is not a problem in warm or moderate climates, but very cold climates require a vapor barrier or vapor retarder to prevent condensation of cold winter surfaces. You can obtain a free copy of a White Paper titled "Proper Design of HVAC Systems for Spray Foam Homes" covering all aspects of building a home with spray foam at bpchome performance.com. This report provides expert guidance for you, your builder and your HVAC contractor on equipment selection, design, ventilation, duct design, indoor air quality, moisture control and other topics.

Windows: If we go back to the design strategies discussed in Chapter 3, you'll recall that windows serve multiple purposes. This means using different styles and ratings of windows for different sides of the home to provide passive ventilation, control the amount of solar heat gain desired, and provide daylighting and views of the outdoors that can improve our mood and contribute to our overall health. Windows usually comprise 10–25 percent of the exterior wall area of a home and generate 25–50 percent of the heating and cooling loads on the mechanical systems.[15]

Window products themselves have come a long way in terms of thermal performance over the past fifteen years, so we now have a wide selection of high-tech glazing products available to us. Known as Low-E or low-emissivity glazing, these windows far outperform the clear glass windows of yesterday. Low-E glazing applications for cooling-dominated climates are tuned to reject solar heat gain, thus keeping the home cooler while allowing visible light to enter. They do this by selectively filtering out the invisible, but heat-rich, infrared part of the spectrum. Different low-E glazing applications for heating-dominated climates allow solar heat to enter but then reflect it back into the home when it tries to leave again. Thus, they act much like a greenhouse glass, helping to heat the home for free. Guidance is available to help you select the best glazing products for your climate and give you insight into how much each option will save you on your annual energy bills. Enter your zip code at the US Department of Energy's Efficient Windows Collaborative website (efficientwindows.org) and it will direct you to window specifications right for your climate and give you insight into how much each option will save you on your annual energy bills.

Beyond coating options, there are numerous choices in materials and performance available, with double and triple pane that help to manage heat transfer and frames that are thermally broken metal, vinyl, fiberglass and, of course, wood and wood composites. The National Fenestration Rating Council (NFRC) is the independent industry- and government-supported non-profit third party that tests and rates windows and doors for performance criteria. National building codes require NFRC labeling. If no NFRC product label is available, the builder must use performance values found in a default table and not those provided by the manufacturer in calculating overall building assembly code compliance. Our advice is that you do not accept any windows that are not tested and labeled by the NFRC.

There are four ratings reflected on the NFRC window and door label:[16] U-factor (also known as U-value), SHGC (solar heat gain coefficient), VT (visible light transmittance) and AL (air leakage). In building performance, the first two are of primary interest, but what you are looking for may vary not only in terms of your climate, but also the location and orientation of the window on your home.

The U-value indicates the thermal conductivity of an assembly in BTUs per hour per square foot per degree of temperature difference, indicating how well the assembly insulates and prevents heat transfer. The lower the U-factor, the better its insulating value. It's common now to find windows with U-factors from 0.27 to 0.40. However, it's important to put that in perspective by understanding that the U-value is the inverse of the product's R-value.

A wall assembly with an exterior cladding, air space, house wrap, insulative sheathing, structural framing, insulation and drywall may have a combined R-value of R-15. The inverse of R-15 is 0.066, which represents the wall's U-value. Windows with a U-value of 0.20, then, have an R-value of only R-5. Even the best windows on the market still perform around three times worse than the average wall assembly, allowing far more heat flow than the same area of wall does. The more windows we put into a wall, the more we are reducing the overall energy performance and comfort that wall delivers. Smaller, fewer and better windows pay off in terms of improved overall building performance, comfort, durability and efficient operations. Remember, designs that strategically place windows for utility, not just as aesthetic embellishments, result in smaller mechanical systems, thereby lowering our initial costs of both windows and HVAC, as well as our operational costs over time.

The Solar Heat Gain Coefficient (SHGC) measures how well the window blocks the sun's radiant heat. Since the SHGC is the percentage of solar heat that comes through, the lower a window's solar heat gain coefficient, the less solar heat it transmits into the house. In a hot climate, you want to reduce solar heat gain, while in colder climates you will want to maximize it. In general, southern builders will try to attain the lowest possible SHGC. Builders in the north are generally more interested in windows that insulate best and select the lowest possible U-factor. In other words, in the south you want to keep the sun's heat out, while in the north you want to let the sun's free heat in and keep the cold out.

If the window is located on a wall where you want solar heat gain in the winter, you may not want a low SHGC rating. In passive solar strategies, windows located on the south side of the house can be protected by overhangs that shade the glass from radiant gain in the summer, but allow that gain in the winter when the sun is lower on the horizon. When using this strategy, you would select a higher SHGC for your solar collection windows to let the heat in. On the north side of the house, there is never direct radiant gain on these windows, so this rating is not as important.

Visible Transmittance (VT) measures how much light comes through a window product. VT is expressed as a number between 0 and 1. The higher the VT, the more light is transmitted through the glass and the brighter your rooms will be. Air Leakage (AL) is indicated by an air leakage rating of the window assembly, expressed as the equivalent of the number of cubic feet of air passing through a square foot of window area when under a given test pressure (cfm/square feet). The lower the AL, the less air will pass through cracks in the window frame.

Windows come in a number of operable styles, including single-hung and double-hung windows, horizontal windows (sliders), casements, louvers and awnings. A good green architect should specify window types for each location in the home to first maximize their particular benefits and then determine how those windows fit into the aesthetic architectural design.

Transom and clerestory windows can be fixed or operable and provide daylighting when installed high on exterior walls or in a cupola or monitor. When operable, they also serve to exhaust rising warm air, as an integral component in a stack ventilation strategy. Double-hung windows (both the lower and upper window sashes can be opened) give you the benefit of cross-ventilation without compromising security. Single-hung windows (only the bottom window sash can be opened) and horizontal sliders (one panel slides either left or right instead of up or down) are less expensive, and can be mixed and matched to serve various room configurations. Casements crank open and closed with the glass panes opening vertically, or rotating sideways either to the outside or to the inside. So if you plan them to crank open in the direction that the prevailing breeze is flowing, these function well to capture and direct that breeze into the living space. Hoppers and awning windows work well in basements and can provide ventilation when used over interior doorways. Awning windows crank open with the glass panes raising and lowering horizontally. This works exceptionally well to minimize water intrusion in case it rains while the windows are open. Casement and hopper window units often have a higher air leakage rate due to the large number of edges that must be sealed to make them tight, while sliding window types have a higher failure rate over time as their gaskets wear out. Check and compare their air leakage rate on the NFRC label to know for sure.

Which materials the windows are made from is a matter of your budget and personal preference. Fiberglass and wood-clad windows are a good choice, if you can afford them, as they are very durable, have long life expectancy and are repairable. Vinyl windows are readily available and dominate the window replacement market. Metal windows should only be considered if they have a thermal break, meaning there is an air space or insulation filling a gap between the inside and outside of the frame, reducing heat transfer. New phase change technologies are making it possible to develop windows that can darken to provide privacy when needed and lighten to increase daylighting. Other technologies may allow windows and other wall and even roof assemblies to reflect or absorb solar radiant heat, as desired for different climates, seasons or times of day, or to act as solar collectors (building-integrated solar photovoltaic). These and other smart home technologies are expected to be on the market within this decade. The future is near!

Roofing: The most important functions of your roof are to provide shelter for your structure and to contribute to optimizing the building's thermal performance. Shelter is more a function of design and installation quality, so it is important to make sure that your roofing contractor uses good-quality flashing materials and methods. Remember, water runs downhill, so all roofing materials, including flashings, should be installed in shingle fashion, overlapping top pieces over bottom ones.

In selecting roofing materials, first go for durability and then think about how you can use the material to impact heat gains or losses, depending on your climate. If you live in a cold climate, you may want a dark-colored roof in order to capture the sun's warmth in the winter. In hot climates, you'll want the opposite and should select materials that are able to reflect back the sun's heat and release any absorbed heat quickly. An independent third party, the Cool Roof Rating Council (CRRC), provides testing and reporting of roofing materials, but does not provide any certification or guidance for recommending materials. ENERGY STAR, on the other hand, does certify and label products according to their SRI (solar reflectivity index) and emissivity ratings.

Historically, white and bright metal roofs have the lowest SRI ratings and dark-colored asphalt shingles (or black tar roofs) have the highest.[17] However, there are new pigments[18] on the market that are coated to be able to reflect sunlight, thereby reducing heat absorption. It is important to look for the ENERGY STAR label for guidance on which roofing materials offer the best rated performance. Not only do dark colors add to the heat load of the home, they also contribute to localized temperature change—the urban heat island effect. As well, the greater the slope of the roof, the more the material contributes to this effect.

In hot climates, ventilating the roof assembly or using a radiant barrier can also reduce the amount of heat conducted through to the living space. This can be accomplished by using furring strips on top of the roof decking, allowing for ventilation under metal roofing. This eliminates condensation concerns and reduces thermal bridging through the roof assembly.

Metal and tile roof products are extremely durable with long life expectancies. If you are considering metal, make certain it is heavy gauge, and if you can afford it, chose a standing seam or concealed fastener system over a screw-down type. Every screw that penetrates your roof decking is a future source of roof leaks, and over time the screws will loosen as the metal expands and contracts with temperature changes. Tile roofs have a long life expectancy.

Tornados, hurricanes or just strong storm winds can do major roof damage. The type of roofing materials you select initially can determine how often you must repair or replace your roof, even if it is only due to normal wear and tear. Metal may seem out of your budget, but if you think that the typical lifespan of a metal roof is three times that of shingles, it is only nominally more expensive. The average composition roof shingle might last 15 years with no major risk events, but you would be considered lucky to go 30 years without an event that is considered a typical risk for your area. So, over the course of 30 years, it would not be unreasonable to have to replace that shingle roof at least two or three times if risk events occur. On the other hand, you could have installed a metal roof that would have survived all of the events of that period and still be providing a durable shield for the building for years to come. Many home insurance firms offer a reduction in annual premiums for metal roofs since they don't need to be replaced with every hail storm and they resist fire. Metal is also infinitely recyclable, so even at the end of its life, it has salvage value.

Another aspect to consider is how the roof rafters are attached to the house. There have been several instances after tornados and hurricanes where a few homes remain almost fully intact in the middle of complete devastation in the surrounding neighborhood. Investigation into the construction methods used in these homes revealed that they shared one key construction detail: the roof rafters were attached to the top plate of the walls with metal reinforcement made for that purpose. Subsequent research has shown that if you can keep the roof on the house, the walls can withstand tremendous wind loads, but once the roof goes, the walls are doomed to fall in. If you are building in an area where tornados or hurricane-like winds occur, it would be wise and a fairly low-cost option to use reinforcing clips, straps or brackets.[19] Hurricane ties can also reduce your insurance costs by as much as 30 percent a year in many places.

You should also think about the best type of roof if you plan to install a solar photovoltaic system for onsite power generation. You want a roof that will stand the test of time, as each time you must replace the roof you will also have to remove the solar system and all of its support structure and replace them after the new roof is installed. That can get quite expensive, and probably will not be covered by your homeowner's insurance if the roof replacement is due to storm damage.

Of course, a very green alternative is a living green roof. This type of structure has been used for centuries around the globe, but has only recently seen some resurgence in interest, as we continue to embrace passive strategies in order to reduce our dependence on mechanical systems. Designed correctly and installed with a good membrane and the appropriate planting medium and plant species for your climate, these systems act as insulation, providing excellent thermal performance. They also mitigate the overall impact on the home's footprint for stormwater management, carbon sequestration and heat island effect.

Interior Finish Materials

The priority considerations for all interior finish products are durability and impact on indoor air quality. It is worthwhile to select styles that are aesthetically pleasing and classic. It's much greener to redecorate with a new coat of paint on the walls than to replace cabinetry, flooring or countertops because they are out of style, yet still functionally sound. For impact on indoor air quality, remember it's not just the product, but also the materials used to construct and install it, including the binders and adhesives (more on this in Chapter 7).

Flooring: Hard-surface flooring materials have long life expectancy and are easier to keep clean than carpet. If possible choose local, salvaged or sustainably sourced wood species, nailed down and finished with natural oils. For your ground-level floor, consider staining and finishing the concrete foundation. If you are using a natural building system, consider an earth floor—they are so comfortable to walk on.

Use natural materials like ceramic tile, materials that are rapidly renewable like bamboo and cork, or materials like linoleum that are made from renewable sources. Have them finished offsite and installed using minimum adhesives (i.e., floating floors that are only glued around the perimeter, not under the entire floor area). Cork makes a beautiful flooring material; it is very easy on the feet and back, has great acoustic control properties, comes in a wide range of colors and patterns and is renewable. Porcelain ceramic tile has color throughout, so if it gets scratched or chipped over time, these blemishes are less noticeable.

Don't take sustainability and durability for granted; make certain that raw materials used to make the products are select grade. Many inexpensive products, like some bamboo flooring, are now coming from sources that harvest crops prematurely. This "green" bamboo does not have the durability and long life that you would expect from hardened mature harvests.

If you must install carpet, make certain it has recycled content or is made from rapidly renewable materials (such as wool or sisal) by seeking products that carry the Carpet and Rug Institute (CRI) label. Also, consider installing carpet tiles, so worn areas can be replaced without having to replace the entire floor.

Cabinets, Countertops and Trim: There are many alternative products on the market made from natural or recycled materials, including cabinets, countertops and trim made from strawboard, bamboo and even sunflower seed hulls. Make sure the binders used in the materials and finishes are low toxicity. Cabinets certified by the California Air Resource Board (CARB) or European Standard (E-1) meet indoor air quality standards.

Solid-surface countertops offer the best durability. There are numerous choices, including natural granite or stone and recycled products, such as glass, paper (yes, we said paper!) and porcelain. Some man-made solid surfaces, such a quartz composite, have antibacterial properties and low waste. Laminate countertops are usually made with urea-formaldehyde binders, although alternatives are now available.

Millwork trim is available manufactured from medium density fiberboard (MDF), strawboard or finger-jointed wood. There may be even greener alternatives available, depending on where you live, including natural local wood species, or agriboard products, made from agriculture waste (like straw from cereal grain production). Of course, the absolute best choice would be no trim, and this fits well in many modern designs.

Finishes, Adhesives and Other Interior Products: Caulks, sealants, adhesives, paints, stains, varnishes and other finishes are big sources of volatile organic compounds (VOCs) that off-gas toxins impacting indoor air quality (see Chapter 7). You can, though, choose to use all-natural paints, which reduce toxins, have low-VOC or no-VOC off-gassing, and clean up with soap and water.

Also, it's best to paint exteriors with lighter colors to reduce heat absorbed into the structure in warm climates, or dark colors to help absorb the heat in cold climates.

Do not assume that one size fits all in caulking and sealant products. There are specialty products for every application that have been developed to provide the airtight seal that high-performance homes are known for. Your building specifications should include brands and product names for each different application. It would be a crime to have a home that costs hundreds of thousands of dollars to build experience building failures (e.g., water leaks, pest or wind damage) due to not spending a couple of dollars on the right product for prevention.

How We Use Building Materials

Sometimes it is not the material itself, but how we use it that makes it green. There are many ways to use common building materials to make your project a darker shade of green. One of the best sources out there for climate-specific building science information is the Energy and Environmental Building Alliance (EEBA: eeba.org). This organization continually researches and publishes information on moisture management (from both exterior and interior sources), air barriers, vapor barriers and just plain good applied building science-based construction techniques. The goal is not only to use the right products, but to use them in such a way as to support the building-as-a-system approach that integrates assemblies in high-performance homes.

Advanced Framing

Optimal Value Engineered (OVE) or Advanced Framing[20] techniques are framing techniques that use less lumber than conventional framing, yet are just as structurally sound. Using 5 to 10 percent less lumber is cheaper and faster because it uses 30 percent fewer framing pieces,[21] which equates to a direct reduction in lumber package costs. But, as important, these methods also present one of our best opportunities to improve the thermal performance of the structure.

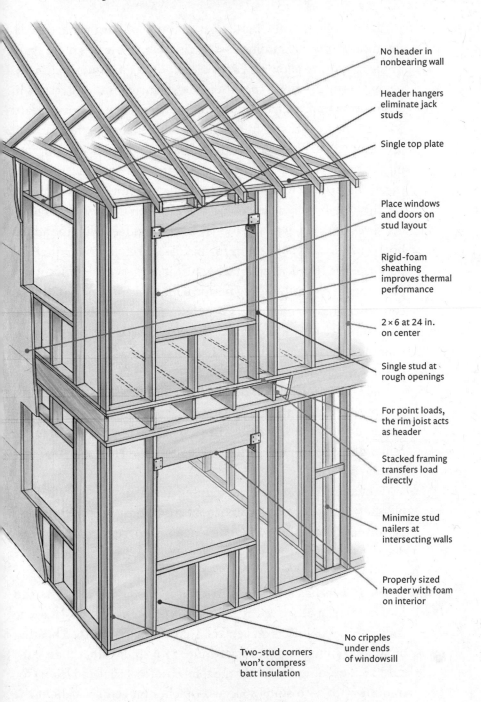

No header in nonbearing wall

Header hangers eliminate jack studs

Single top plate

Place windows and doors on stud layout

Rigid-foam sheathing improves thermal performance

2 × 6 at 24 in. on center

Single stud at rough openings

For point loads, the rim joist acts as header

Stacked framing transfers load directly

Minimize stud nailers at intersecting walls

Properly sized header with foam on interior

No cripples under ends of windowsill

Two-stud corners won't compress batt insulation

FIGURE 4.2. Advanced Framing techniques. Credit: *Fine Homebuilding*.

Remember, thermal performance is one of our primary goals. Because Advanced Framing uses less lumber, this leaves more space in the cavities to allow for higher levels of insulation, resulting in better thermal performance. In fact, Advanced Framing results in a 75 percent improvement in thermal performance[22] over standard 2 × 4, 16-inch-on-center framing. The lumber cost savings can add up to enough to cover the additional costs associated with improving the thermal performance, including the cost of adding insulated sheathing to reduce thermal bridging through the wall assembly.

This is definitely a case of "less is more," in that using *less* lumber saves trees and uses *less* money in your budget, with the added bonus of leaving *more* room for insulation, giving *more* thermal performance. Think about that because it represents significant improvement in your building performance for minimal additional construction costs *and* lowers your long-term utility bills. So if you are planning a framed structure, you should make certain that your framing contractor has been trained in and practices the methods discussed here.

Many computer-aided design (CAD) programs can be set to a grid of either 16 inches or 24 inches on center (the distance from the center of one framing member like a wall stud or ceiling joist to the center of the next) to allow for ease of designing to these two basic spacing criteria.

Providing these Advanced Framing details in your architectural plan sets can assure that you benefit from those cost savings and performance benefits. Better efficiency, improved comfort, and reduced costs are all achieved in each of the following methods:

Framing 24 Inches on Center: Exterior wall studs, floor joists and roof rafters can be spaced at 24 inches on center (as opposed to the conventional 16 inches on center). Depending on the load bearing on walls, framing lumber may require 2 × 6 studs rather than standard 2 × 4 framing. Note that the total cost (material and labor) for framing with 2 × 6 studs spaced 24 inches on center is about the same (since 30 percent fewer studs and only a single top plate are

required) and often less than what it would have cost for 2 × 4 studs spaced 16 inches on center. Because there are fewer studs to cut, there is less waste. This also saves labor for both your electrical and plumbing contractors, who now have to drill fewer penetrations for mechanical runs.

However, with most types of cavity-fill insulation (depending on climate R-value requirements), it may cost more to fill a 2 × 6 cavity than to fill the same structure framed with 2 × 4s. This is not only due to the increased depth of the studs but also to using less framing materials overall, so it will take more insulation material than it would have for the same depth, regardless. In addition, the added two inches of wall thickness will require extension jambs at all of the windows unless drywall returns are used.

In-line Framing: Aligning the floor, wall and roof framing members directly above one another so the loads are transferred directly downward, requiring no additional structural support, can save considerably on structural engineering and framing costs. With in-line framing for improved load stability, double top plates can be eliminated because the load is distributed evenly through the remaining single top plate. Note that studs that are 24 inches on center are placed in direct alignment with floor joists spaced 24 inches on center and directly below roof trusses spaced 24 inches on center. The structural concept is to align all point loads to carry the weight directly down to the ground.

Headers Sized for Actual Loads: Structural headers are often over-sized or installed over all window and door openings, regardless of whether or not they are structurally necessary. When the size of the window used is specified in conjunction with in-line framing, headers are not necessary because no studs need to be cut. If walls are not load bearing, no headers are required over window or door openings. Having your structural engineer specify which areas will require headers, as well as the size of each header required, will save both materials and money.

In most cases right-sized headers can be pushed to the outside of the framed wall assembly, allowing for insulation on the inside of each header cavity, which not only improves the overall thermal performance of the wall assembly but also eliminates thermal bridging at the headers. Note that it is possible and now required by code to insulate headers by using foam sheathing as a spacer in place of plywood or oriented strand board (OSB), either between or on one side (preferably the exterior side) of doubled headers. This technique uses scrap foam sheathing to reduce thermal bridging through the wood header.

Two-stud Corners (California Corner) with Drywall Clips: This method of corner framing uses only two studs, saving material and providing space for additional insulation in the corner. To attach drywall in a two-stud corner, drywall clips are fitted onto the edges of the drywall before being attached to wood or steel studs. This eliminates the need for an additional stud in the corner to attach the drywall.

Window and Door Placement: By aligning at least one side of each window and door to an existing wall stud, use of an additional jack stud is not necessary. If the window or door width does not completely fill the cavity and align with the next stud, you can attach the other side to the next stud with a metal hanger. This eliminates the need to frame additional studs to support the load transfer around these penetrations in the wall assemblies.

Interior Partition Walls Intersecting with Exterior Walls (T-walls): Traditional framing addresses T-wall intersections by adding studs at each side of the partition solely for the purpose of providing a surface for attaching drywall. Ladder blocking between the exterior studs behind the partition wall uses two-foot scraps of lumber to provide the same supporting structure and allows for much better wall insulation and reducing thermal bridging. You can use scrap wood for ladder blocking, reducing the additional lumber you need to purchase.

Exterior Wall Assembly

Continuous and contiguous air, thermal and moisture barriers are essential to healthy, high-performance homes, but it takes meticulous detail work to achieve this. A combination of rain screens, caulks and sealants, flashings and weatherstripping is required. Every penetration, transition and margin of the building must be addressed. Below are some general guidelines for exterior wall construction not only for long-term durability but also to achieve the performance that you expect from the product.

Thermal Barriers: "Thermal barrier" is a fancy term for insulation. In a high-performance home, insulation should be installed on all exterior surfaces in an unbroken sequence. Any gaps, voids or breaks in the insulation coverage of the entire building assembly can result in heat loss or gain. This is shown in the image below with the use of a thermal imaging camera. In a color photo, heat loss shows up as warm yellow or orange and cool well-insulated areas are blue or black. In this black-and-white rendition, light and bright areas indicate heat loss.

FIGURE 4.3. This thermal image shows heat loss through poorly insulated areas of the home. Credit: The Renewables RouteMap Team, The Scotland Government, UK.

Thermal bridging is the rapid transfer of heat through a building component when that component has less thermal resistance
(R-value) than materials surrounding it. Framing materials offer a
good example of thermal bridging through the building envelope.
Wood has an R-value of a little less than one per inch, so a typical
2 × 4 stud has an R-value of around 3.5. Compared to the surrounding insulated wall cavities, if perfectly installed to manufacturers'
specifications to achieve R-13 or R-19, that's quite a difference. So, if
you look closely at the image above, the thermal movement through
the wood-framing members allows you to see all of the studs and
even the roof rafters glowing with the heat they are losing. Thermal
bridging can greatly reduce the effective insulation value of a wall,
floor or ceiling.

Another place where poorly insulated wood is typically used is
for structural headers to displace the vertical loads over windows
and doors, as was mentioned previously. This volume of uninsulated wood creates large areas of thermal bridging, significantly reducing the overall thermal performance of the entire wall assembly.
As was mentioned earlier, it's unfortunate that many framing crews
are taught to install headers over every window and door, even
when they are located in non-load-bearing walls. This usually happens due to a lack of framing detail provided to them by the structural engineer or truss designer. Best practices are to install headers
only where they are required structurally, to size them only for the
actual load they are to carry and to insulate them. Many insulated
structural header products are available on the market, or you can
make your own by sandwiching a rigid foam board panel between
two layers of wood (or structural wood product) to create a thermal
break.

By adding a layer of rigid board insulation on the exterior of
our entire wall assembly, we can reduce or eliminate thermal bridging, as this material provides an insulated break between the wood
framing and the exterior heat source. By sealing the attic and insulating over the exposed roof rafters, we can reduce or eliminate
thermal bridging there using the same approach. We could also

choose to use an alternative building system, like SIPs, ICFs, AAC block or a natural material, which could significantly reduce thermal bridging in the building assemblies.

In Chapter 3, we recommended raised heel or energy truss design as a remedy for insulation gaps between the top of the wall assembly and the edge of the roof assembly. It is evident from Figure 4.3 that this home suffers from poor insulation in the soffit area, a significant source of heat loss in the winter. These are like holes in the thermal envelope, and so the walls perform as if someone has left a window or door open, putting additional strain on the air conditioner or heater as it attempts to provide comfort under these conditions.

Finally, note the heat loss through the foundation or basement perimeter. This has become a more important issue as we built tighter thermal envelopes, which should enable us to reduce the size of air conditioning and heating systems required to keep them comfortable. However, this heat loss through the foundation assembly can result in raising the heating loads, negating any savings achieved in the main building assembly. In fact, we have seen instances in the last couple of years where heat pump system sizing is being determined by these heat losses, driving up heating loads even in cooling-dominated climates with very mild winters. This means that although we did a good job reducing the cooling loads through building science and envelope improvements, we were forced to install a larger HVAC heat pump system to handle the heat loss through the foundation in the few very cold days of winter that occur. This is the best argument for insulated slabs in any location that has any chilly winter days. An insulated slab or basement can reduce the heating load on the home by as much as 25 percent or more[23] depending on your climate and house plan.

Air Barriers: Also, since insulation is a material full of air pockets, it is important to stop airflow through that material. For insulation to be effective, it must be encased by an air barrier. Air barriers function to keep air from freely flowing through insulation, allowing it to achieve the thermal performance (R-value) at which it was rated. To be effective, insulation and air barriers should be both continuous and contiguous, meaning that every exterior building assembly is insulated and encased by an air barrier. Research has proven that installing insulation without an effective air barrier results in a huge reduction in the effectiveness of the insulation and high bills with poor comfort.

Wherever the insulation is installed, there must be an air barrier *in contact* with the insulation on all six sides, leaving no insulation exposed; this prevents convection currents. The wall studs, along with the top and bottom plates, close up four sides. The exterior sheathing encloses the outside of the cavity, and drywall *normally* encloses the inside, but not without exceptions.

These exceptions are because there are areas of the thermal envelope that may be insulated but often do not have drywall installed on the inside of the cavity. This includes fireplace and HVAC chases and behind bathtubs when these features are located on an exterior wall. This can also include a stairwell on an exterior wall, even if part of the area under the stairs is a closet. Usually the under-stair closet ceiling slopes down to a point such that the bottom few steps of the stair would create a ceiling height too low to be usable. These bottom few steps, if on an exterior wall, will usually not have that area of the wall enclosed with drywall. In these areas, it is necessary to install some other type of air barrier to encase the insulation on the inside of the wall assembly.

Note that an air barrier is shown installed on the inside and outside of the wall common with the attic space, often called a knee-wall or pony wall. These are vertical walls that separate a room from an attic space. Typically, builders do not install an air barrier on the attic side of these walls. They just stuff some batts into the cavities and call it good enough. They also don't place air blocking in the big holes under the knee walls where the ceiling framing runs. This leaves dozens of big holes (16 inches by 8 inches) open so that outside attic air easily blows between the uninsulated floors and ceilings. In cold climates, the result is often frozen pipes between the

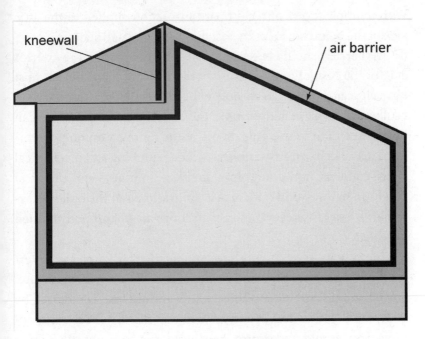

FIGURE 4.4. Air barrier. Credit: Adapted from ENERGY STAR.

floors of the home where you would think that cold air shouldn't be able to go. Very often rooms over garages are uncomfortable because they suffer from both of these problems. These areas, even if insulated, are large holes in your thermal envelope when not sealed by some type of air barrier.

The exception to the air barrier installation requirement is if you are installing blown insulation on the attic floor. For blown-in insulation in the attic floor, significantly higher R-values are typically required by building codes to achieve the desired resistance to heat needed here. Since the insulation is not installed vertically, it is not as susceptible to convection loops and for this reason doesn't need to be encased on the sixth side. The depth markers that are commonly seen in this type of installation ensure the depth of the insulation achieves its stated R-value.

A common hole in attic insulation occurs at the location of an attic scuttle hole or attic stair. Any attic access that penetrates the thermal envelope should be well sealed with weather stripping, with multiple layers of rigid board insulation applied to the attic side of the board cover, or by installing an insulated stair unit.

Air barriers should be sealed at all penetrations. On the exterior side of the insulation, the house wrap or rigid foam board must have all seams taped. To complete the air barrier, it is necessary to caulk and seal all penetrations in the building envelope. There can be no exception to this rule. Some of the more common penetrations in wall and roof assemblies include plumbing and mechanical vents, condensation drain pipes, fireplace chimneys and electrical conduits, fixtures, and outlets. Air infiltration into the building assembly occurs wherever these penetrations are not properly flashed or sealed.

The way to know if you have an effective air barrier is to test the house under pressure and then measure the air infiltration rate. Since 2009 this test, often called a blower door test, is required by code, but if you do not live in an area that mandates code inspections, you should make sure your builder is aware of the blower door test and that you see the results. The house must be tested and proven to be a very tightly sealed structure by achieving no more than 5 air changes per hour at 50 pascals of pressure (ACH50). This standard will be restricted even further to no more than 3 ACH50 in climate zones 3–8 by the 2015 International Energy Conservation Code (IECC) (advance information June 2014).

Ice Damming: Ice dams are a source of tens of millions of dollars in home damages each year. Homeowners have tried heat tapes at the edge of the roof, rakes, more ventilation in the attic and a million other ideas to stop this threat, but with limited success. The building science community has been able to determine how and why ice dams occur, and this has led to a successful strategy to stop them. They found that the problem isn't at the eave—that's just where it becomes evident and does its damage. The ice dam itself is just a symptom of the real problem.

Preventing ice dams is all about creating an effective air and insulation barrier at the ceiling of the house. An ice dam starts when heat is allowed to rise from the ceiling of the house at what building science calls thermal bypasses (explained below) and melts the underside of the snow pack on the roof. This meltwater runs down the warm roof as liquid at just above freezing. It stays liquid until it reaches the exposed eaves of the house, and there the temperature of the roof deck suddenly drops because the deck is fully exposed on the bottom side to the cold outside air. The meltwater flash freezes at the eave, and the ice dam begins to form. As more water melts and runs down the roof, it builds up a small lake behind the ice dam and backs up under the shingles to run down the walls of the home. If the snow pack stays frozen, there can be no ice dam. The solution is to correct the cause, not to deal with the ice dam itself. In other words, stop the heat from rising at the ceiling of the house and the snow pack will stay frozen and no ice dam will be able to form in the first place.

This requires us to seal all thermal bypasses and insulation defects in the ceiling of the home. A thermal bypass is a place where the insulation and ceiling air barrier have holes in one or both system (remember: "continuous and contiguous"). These are places like open utility chases for plumbing, wiring and ducts, dropped ceilings and other irregularities that provide opportunities for breaks in the drywall air barrier and insulation at the ceiling. These openings will act as a chimney for warm air to escape into the attic where they rise and warm the roof deck.

All thermal bypasses must be sealed airtight with rigid materials like plywood or foam board, then air sealed with caulk or foam sealant and then covered thoroughly with insulation. This is a job best done by the framing crew during the initial framing of the home. Subcontractors who later penetrate these air barriers must be held responsible for resealing them once they have installed the plumbing, wires or ductwork that runs through them. Any areas of missing insulation must be fully insulated.

Recessed can lights are notorious thermal bypasses. They are not only ventilated to allow warm house air to pass through, but they generate their own heat when the lights are on. The ENERGY

STAR Certified Homes Program has an excellent list called the Thermal Bypass Checklist[24] showing everything that must be sealed for a home to be certified.

The best solution in new construction is to seal everything on the Thermal Bypass Checklist and to use only recessed light fixtures that are airtight rated and also rated for insulation coverage. These lights are often called AT/IC-rated recessed fixtures. They should meet ASTM E-283[25] and be labeled as such.

If you can't replace the old leaky can lights in your existing home, consider a code-approved option. Build a sealed box out of drywall that leaves a clearance around the fixture as specified by the manufacturer. Cover the cans with the drywall boxes and seal the boxes down to the ceiling drywall. This will stop the air from rising through the cans and heating the roof deck and starting the process that leads to ice dams.

Radiant Barrier: A radiant barrier is the foil-faced roof decking (with the foil facing the attic side) that stops radiant heat gain through a vented roof assembly in hot climates. This is usually one of the first upgrades for existing construction if your roof design allows good access for installation, along with improving the insulation in the attic if needed. Note that radiant barriers require an open, ventilated air space on the side facing the house, so they should not be installed in a sealed attic, as they simply don't work when foam insulation is in contact with them.

If you live in a cooling-dominated climate, you should also strongly consider using aluminum foil-faced radiant barrier roof decking material with an emissivity of 0.05 or less[26] to keep your attic cooler in the summer. This will at least prevent much of the radiant heat gain through the roof assembly. This results in helping to take some of the heat load off your air conditioner and ducts during hot weather.

Moisture Management: Water is the enemy of all building materials, and can easily destroy everything we do to make our home healthy, safe, comfortable and durable. That may sound a bit over the top, but it's the plain truth. Water can come from outside in the form of rain, or it can come from inside in the form of humidity that condenses on a cold surface. Either way, the materials are now unhealthy waste products.

The best way to prevent moisture issues is to ensure that your home has an effective raincoat, or drainage plane as building science often calls it. This is achieved by the proper layering of water-proof materials in the wall assembly. Starting with the top of the house, the roof, you should be able to follow the flow of water down and off or out of the building assembly and verify that is a clear, unobstructed direct path.

To do this, we'll begin at the roof in order to understand how shingles work to keep water out of your home. They are always laid from the bottom up so that the upper shingles overlap the bottom shingles. Remember, water runs downhill. As the water runs down the roof, it would have to run uphill to get under the next shingle. Since this is very unlikely, the overlapping layers form a water-resistant system.

Moving from the roof to the walls, all the other waterproofing materials, like house wrap and metal flashing, must be installed from the bottom up, overlapped the same way roof shingles are laid. This way water flows across the top of the layer and has no way to get under and inside the flashing. This sounds simple, but it is very common to see these materials installed backwards! Too many crews today are not well trained. They only know that one layer must go under another, but reverse the order. When this is done, the water has a clear path inside the waterproofing. In fact, the lower layers act as an opening directly into the house, catching the water like a gutter.

Moisture management details mean the difference between living with years of worry-free enjoyment of your home and major, expensive repair costs. To make sure it's done correctly, you need to understand the function of each material used. Cladding is the material that forms the exterior façade of your structure and provides ultraviolet (UV) light protection for the materials inside the wall assembly that would otherwise be degraded by direct exposure. Some green cladding choices include natural materials, like stone, and man-made materials that are durable, like cement-based siding. It is important to understand that exterior cladding materials are intended only to shed bulk water during heavy rains, not prevent water penetration. Stone, brick and mortar are porous and water will soak through them in a matter of only a few (15–30) seconds![27] Wood, vinyl and fiber-cement siding all readily leak. Since all claddings and windows leak, it is the responsibility of those building homes to take the steps necessary to manage the water that penetrates them.

The most aggressive water management system requires a clear one-half to one-inch air space between the cladding and the drainage plane, with weep holes at the bottom of the wall to allow rain-driven moisture to escape. This type of drainage system is more commonly referred to as a rainscreen. To assure a consistent, unobstructed air space, 1×4 furring strips are installed, which allow for gravity drainage. The drainage plane material is typically a house wrap (with a 25-year warranty) and/or asphalt-impregnated builder's felt meant to keep liquid water out of the wall assembly.

One of the important components of the rainscreen is house wrap, which is a very interesting material. These wraps are engineered to be highly resistant to penetration by liquid water, but highly permeable to water vapor/humidity. This is so that they can do both of the jobs we need them for. First, they keep the walls dry by stopping liquid water like rain from getting to the wood. Second, they allow the walls to dry out once water has penetrated past them by allowing water vapor to pass through.

Remember, every wall cladding system and every window eventually leaks so we must plan for this. While most house wraps can hold out liquid water for a long time, they have water vapor permeability ratings from 7 perms up to 58 perms. The perm rating tells us how quickly water vapor will pass through a material. The perm ratings of house wraps are all high enough to allow for wet materials to dry out by releasing the water vapor from the wall system. The secret to how house wraps do this is that they control the size of the holes in the material. At a microscopic level, they have holes big enough to allow individual water vapor molecules to pass through, so any moisture inside will eventually dry out. But when water vapor molecules bunch up to make a molecule of liquid water, they are then too big to fit through, preventing liquid water from entering the building assembly through them.

If you are using stucco as your cladding, leaving a half-inch air gap behind that material is usually not possible, so the requirement for the water drainage plane is for a double layer of water-resistant underlayment. This double layer can be comprised of two layers of house wrap or asphalt-impregnated builder's felt, a layer of either of these and a layer of paper-backed lath or paper-backed lath over a layer of rigid foam sheathing with the seams sealed with approved tape. The first layer serves as a bond break for the second layer where the real drainage takes place. Two layers of drainage plane behind stucco walls is now a code mandate and the position of the Portland Cement Association.[28]

If you visit the website of the manufacturer of your house wrap or drainage plane product, you will find step-by-step detailed drawings and instructions showing how the builder must flash the windows, doors, porches and other areas of your home vulnerable to water penetration. Be sure that this is done right and don't be afraid to speak up if the crews mess up. If it's not done right, your manufacturer warranty for your house will be voided. You only get one shot at doing this right.

Building science technologies are beginning to develop components that integrate several layers of assemblies into one product. These products are available for both roof and wall assemblies. Some combine continuous foam insulation to stop thermal bridging and an engineered panel that replaces structural wood sheathing in a single nail-applied product. These products incorporate a built-in water-resistive barrier, eliminating house wrap or felt, and are installed and taped at all joints. This provides a continuous structural, insulative air barrier and water drainage plane. These types of engineered, prefabricated and performance-tested multi-purpose systems are available now, and represent the future of high-performance building materials.

Flashings: There's an old saying that fits this topic very well: the devil is in the details. That's because it's how we flash windows, doors, porches, chimneys and other key areas of the exterior walls and roof that will keep the water out of your home. Flashings are one of the most important and overlooked components of the building envelope that make or break long-term resistance to damage. Make certain that your framers and roofers are using the right materials for the job and installing them in the proper fashion. This is not the place to cut money from your budget! Flashings of all sorts are manufactured to fit every joint and intersection of your exterior building envelope, to cover all the different angles, seams, gaps and penetrations. It's not just the materials; even more importantly, how they are installed keeps the water running downhill.

With windows, the process begins with a sill pan or a layer of waterproof material covering the bottom of the rough opening that the window will be installed in. This material should be turned up at the corners at least six inches on the inside of the studs and then extend out on top of (not under) the house wrap below the open-ing. The house wrap has a flap cut above the window. The window is then installed, and the sides are flashed with an adhesive tape, followed by a layer of flashing tape across the top of the window flange. The flap of house wrap is then brought down over the top flange, and it is taped in place with flashing tape. The bottom flange of the window is not flashed. This will allow any water that does get into the opening to drain out and flow down the waterproof drainage plane of the house and out the weep holes or screed at the bottom of the wall. By the way, the order that these flashings are installed, from the bottom up, is critical. For curved windows and doors there are flexible flashing tapes on the market now that work very well.

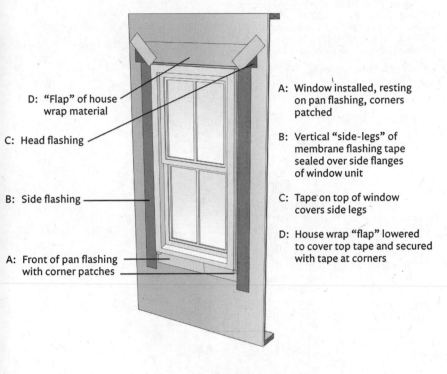

D: "Flap" of house wrap material

C: Head flashing

B: Side flashing

A: Front of pan flashing with corner patches

A: Window installed, resting on pan flashing, corners patched

B: Vertical "side-legs" of membrane flashing tape sealed over side flanges of window unit

C: Tape on top of window covers side legs

D: House wrap "flap" lowered to cover top tape and secured with tape at corners

FIGURE 4.5. Flashing layered application.
Credit: Adapted from drawing provided by Building Science Corp.

The process for porches and decks is much the same. The drainage plane material is fully installed before the ledger board is put in place. After attaching the ledger board, the drainage plane is cut above the ledger board to which the joists will be attached. A flashing material like metal or flexible waterproofing is attached to the sheathing, and wrapped around the front of the ledger board. The drainage plane is then turned down over this flashing and taped in place. You can obtain detailed drawings showing exactly how these openings need to be flashed on the websites of the house wrap or window manufacturer for your project.

The addition of an insulated sheathing to the wall assembly reduces thermal bridging. Or you can replace the rainscreen assembly that we just described with an insulated sheathing panel system that is taped at all seams (with a 50-year warranty). These types of

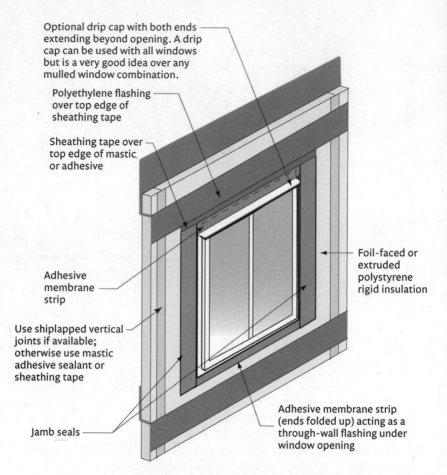

Optional drip cap with both ends extending beyond opening. A drip cap can be used with all windows but is a very good idea over any mulled window combination.

Polyethylene flashing over top edge of sheathing tape

Sheathing tape over top edge of mastic or adhesive

Adhesive membrane strip

Use shiplapped vertical joints if available; otherwise use mastic adhesive sealant or sheathing tape

Jamb seals

Foil-faced or extruded polystyrene rigid insulation

Adhesive membrane strip (ends folded up) acting as a through-wall flashing under window opening

FIGURE 4.6. Insulated sheathing window installation.
Credit: Building Science Corp.

water management systems are always employed behind masonry or stucco walls (because water penetrates those materials so rapidly) and behind wood walls, too, in rainy or marine climates.

Building a Dry, Healthy Basement or Crawlspace: If your plans include a basement or crawlspace, building it to stay dry is a must to ensure a green, durable and healthy home. From a building science and physics perspective, a crawlspace is just a basement with short walls. The physics of moisture, temperature and condensation remain the same. If you build your basement or crawlspace incorrectly, it is very difficult, disruptive and costly to correct the error in later years. Just as we did in building our exterior envelope, building a basement right means that we deal first with liquid water (water in the liquid state can enter through either the floor as rising groundwater or through the walls), next with condensation as a source of water, and finally with capillary water movement.

There has been a great deal of research into the question of how to best build a basement. The Building America Program, the Canadian Mortgage and Housing Corporation (CMHC) and others have completed a multi-year investigation[29] into this question and have concluded that the basement wall that performs the best is one that is damp proofed on the exterior to control capillary action, with rigid foam board installed as a drainage plane on the exterior.[30] By placing the rigid foam board in this location, two benefits are achieved. The first is that the wall is now inside a layer of continuous insulation and thus fully isolated from the cold soil and air. The second is that this means that the wall stays warm and is very unlikely to experience condensation on the inside in either winter or summer months. This makes building a finished basement much easier because it eliminates the worry about warm inside air finding a cold surface and causing problems by condensing.

The second-best option is to install the foam board either in the middle of the wall using a manufactured system like an insulated concrete form (ICF) or on the inside of the wall. Many home-builders prefer to install the rigid foam board or closed-cell spray foam on the inside of the basement wall to protect it from damage during construction. This layer of foam can be XPS, EPS or spray-in-place foam. If you take this approach, be sure that all seams, gaps and edges are fully caulked and sealed so that no inside air can get to the concrete wall, because with this approach, it will be cold and condensation will occur. The seams between sheets, and the top and bottom of each sheet, must be effectively air sealed to keep warm moist inside air from coming into contact with the cold wall. The old-fashioned use of a framed wall with plastic vapor barrier and blanket insulation is no longer considered a viable approach.[31]

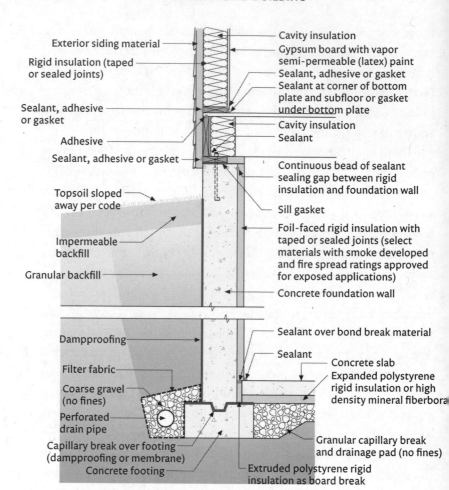

FIGURE 4.7. Basement with interior insulation.
Credit: Building Science Corp.

Concrete is full of millions of tiny pores that inherently lead to very strong capillary action, wicking water into the structure unless that action is controlled. Dampproofing[32] is essential to protect the structure from this. This is often accomplished by using a paint-on or trowel-on material formulated for this purpose.

In much of the country, dealing with groundwater often entails installing a French drain (underground drain) and a sump. The French drain should be located at the bottom of the concrete grade beam supporting the wall and below the level of the basement or crawlspace floor. It should be surrounded with a generous layer of free-draining river rock and the whole assembly wrapped in a geo-textile filter fabric to keep the drain pipe from becoming clogged with soil and to break the hydrostatic pressure of the water pressing against the wall, allowing it to drain down to the French drain. The next step is to dampproof the below-grade portion of the wall and ensure that soil-borne water can drain freely down to the French drain.

Due to the strong capillary action of concrete, you should also be sure that there is a capillary break like a sill seal or 30-pound asphalt-impregnated builder's felt between the poured footing and the wood wall it supports. This also aids in creating an effective air barrier at this uneven location. The control of soil moisture can be aided by ensuring that the outside soil grade is sloped away from the house in all directions with a drop of six inches for every ten feet of horizontal run. Installing gutters and downspouts that take the roof runoff at least three feet out away from the foundation will contribute a great deal to keeping a basement or crawlspace dry.

If you have a concern with radon (see more about radon in Chapter 7), this is the time to place a layer of coarse rock beneath the basement slab and cover that with a taped and sealed layer of ten mil polyethylene. Install a PVC pipe and stub it out so that it is above the slab. Later you will connect a pipe to a riser and run that up through the house and out the roof. An in-line fan rated for continuous duty is installed in this pipe to capture the radon from below the layer of plastic and exhaust it safely above the roof.

Since 2000, the International Building Codes (ICC) have recognized and allowed two methods of crawlspace construction. One is the old ventilated crawlspace. The other is a sealed, unventilated crawlspace. If your local building official needs a code reference, this can be found in Sec. R-408 of the International Residential Code and Section 402.2.10 of the International Energy Conservation Code.

In the past, it has always been common practice to build crawl-spaces with ventilation openings at a rate of one square foot of open area for each one hundred and fifty square feet of crawlspace floor area. As most people who have ventured into a crawlspace will report, they are almost always damp and musty with some evidence of mold or decay on the wood. Research has also shown that even dry soil in a crawlspace will evaporate between 10.2 and 19.1 gallons of water per day. This, along with humidity in outside ventilation air, are the two primary sources for the moisture that makes so many crawlspaces damp and musty smelling.

Some people have thought that adding a powered ventilation fan would make the crawlspace drier. Sadly doing this in most climates actually makes the crawlspace wetter and speeds mold and decay of the wood. Powered ventilation of crawlspaces in areas with humid summers will make them wetter, not drier, contributing to rot and mold growth. The same is true for powered attic ventilation, which doesn't result in the best home performance in any climate. Powered attic ventilation can actually pull conditioned air up from the house below, resulting in higher indoor humidity and negative pressure that can cause backdrafting of carbon monoxide.[33]

With an unventilated crawlspace, the insulation is placed on the walls, not under the floor of the house. This is more energy efficient. The soil is sealed with plastic to control water evaporation. There are no vents. Bulk water is controlled with French drains if necessary. The entire floor (all exposed soil) must be covered with a layer of at least six mil plastic, overlapped at the seams and taped or otherwise sealed. The plastic should also be sealed around all piers and to the exterior walls. Latex water-based HVAC mastic is useful for this application. Finally, mechanically ventilate the space with about 20 CFM of conditioned air for each 1,000 square feet or by installing a small mechanically operated duct or an exhaust fan sized to provide this rate of air exchange or a small dehumidifier set at 50 percent.

Besides moisture control, there are other benefits to a sealed crawlspace whatever your climate zone. For one, the floors of your home will always be at a comfortable temperature. Your pipes will be inside partially conditioned space so they won't freeze in the

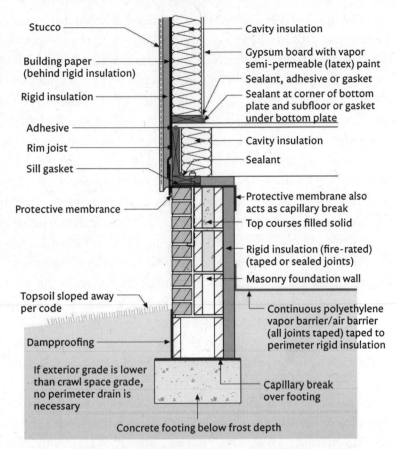

Stucco — Cavity insulation

Building paper (behind rigid insulation) — Gypsum board with vapor semi-permeable (latex) paint

— Sealant, adhesive or gasket

Rigid insulation — Sealant at corner of bottom plate and subfloor or gasket <u>under bottom plate</u>

Adhesive — Cavity insulation

Rim joist —

Sill gasket — Sealant

Protective membrane — Protective membrane also acts as capillary break

— Top courses filled solid

— Rigid insulation (fire-rated) (taped or sealed joints)

— Masonry foundation wall

Topsoil sloped away per code

Dampproofing — Continuous polyethylene vapor barrier/air barrier (all joints taped) taped to perimeter rigid insulation

If exterior grade is lower than crawl space grade, no perimeter drain is necessary

— Capillary break over footing

Concrete footing below frost depth

FIGURE 4.8. Crawlspace with interior insulation.
Credit: Building Science Corp.

winter. Sealed crawlspaces are dry, clean and generally odor free. They are in fact remarkably clean and healthy spaces. We said earlier that a correctly built crawlspace is just a short basement. Now you can see that except for the low ceiling it could be fully habitable!

All crawlspaces would benefit from having the soil covered with a six-mil or thicker plastic vapor barrier as shown in the diagram above. This will stop the soil from contributing so much water vapor to the crawlspace. To quote from the research findings, "Building America research has found that closed, conditioned crawlspaces perform better than vented crawlspaces in most parts of the United States."[34]

There has been extensive research aimed at solving the excess water vapor problem by the building science community over the last 20 years. The Oak Ridge National Laboratory and the Building America Program of the US Department of Energy have published recommendations for crawlspace construction[35] that have been incorporated into our building codes. The problem comes back to climate differences. If you are building in a mild climate where, for the majority of the year, the outdoor air is dry, with a low dew point and relatively low humidity levels, then a ventilated crawlspace can be a good option. On an annualized basis, there will be more days when the wood is dried by the introduction of outside air than when it is made wetter. But in areas where the outside air is humid with a relative humidity over fifty percent much of the year, the net annual impact of introducing that air into the crawlspace will make the wood damp, not dry. The farther south or nearer a coast you live, the more closely you should look at a sealed, unventilated crawlspace. If you live in a cold and dry, or very arid climate, then ventilation can work for you, but your pipes will still be in danger of freezing.

Vapor Barriers: What about moisture in the form of water vapor? Water vapor moisture is produced both outside (humidity) and inside (steam from cooking and cleaning). The second law of thermodynamics says that stuff moves from areas of greater concentration and higher energy to areas of less concentration and lower energy. With regard to vapor, air carrying the moisture (vapor) always moves from high pressure toward low pressure, and water moves from wet toward dry and from warm toward cool. When water vapor hits a cool surface, it condenses and changes from vapor to a liquid. If this is within the wall assembly, that becomes a problem, leading to mold growth and issues with rotting assemblies and poor indoor air quality. Vapor barriers (like vinyl, polyethylene or the asphalt-coated Kraft paper face of our insulation batts) can either help to prevent this from happening, or can actually contribute to the problem, depending on the climate.

So you may or may not need or want a vapor barrier. The goal is to control or stop condensation. There are two ways to do this. One is to stop the warm moist air from coming into contact with the cold surfaces. The other is to warm the surfaces so that they are too warm for condensation to occur.

In the past, we primarily used the idea of stopping the moisture using vapor barriers in our wall assemblies. This is the concept behind the plastic vapor barrier covering the studs. Getting a perfect plastic vapor barrier installation is very hard and detailed work, and there have been far too many instances where small overlooked holes have caused major problems, so now there is another option. With the advent of rigid foam insulation board, we can now warm the wall surfaces to prevent condensation. And this provides the additional benefits of increasing the total insulation of the wall assembly and reducing thermal bridging.

When we do use a vapor barrier, where we place it is determined by which direction the wall will dry towards. Remember, our goal is to stop the water vapor from finding a cold surface and condensing while still allowing the wall to dry in the other direction. Strange as it might seem, northern homes dry *out* while southern homes dry *in*.

For example, if your home is in a very humid, cooling-dominated climate like Dallas, Texas, (or in a mixed-humid climate like in the Midwest), the direction of water vapor drive is from the warm, humid outside air toward the dry, cool air inside of the air conditioned home. As moist air comes in contact with the backside of cool conditioned wall surfaces, condensation and related problems can occur. This is especially true if the owners have kept the house at a temperature below the outdoor dew point temperature. If we had placed the vapor barrier on the inside of an exterior wall, it would have exacerbated the problem by halting the ability of the water vapor to dry to the inside. Many a builder has removed vinyl wallpaper and found the drywall under it covered with mold and not understood the source of the excess moisture. So the right place for the well-sealed vapor barrier/vapor retarder is to the outside of the wall assembly, allowing the moisture to dry to the inside.

Now, let's consider a home in a heating-dominated climate like Minneapolis or Toronto. The warm/humid side of the house walls most of the year is the inside, and the cool/dry air is on the outside of the house. In this climate, the water vapor is driven from the inside toward the outside through the building assemblies in long winter. As this warm, humid air reaches the backside of the cold exterior sheathing, it again causes a condensation problem. The place to put the vapor barrier would be on the inside of the wall assembly.

Except for extremely cold climates, we can skip the vapor barrier entirely and opt for installing the exterior foam sheathing, which keeps the wall assemblies warm enough to prevent condensation. We call this putting a coozie on your house. The thickness of the foam sheathing required depends on your climate zone. In the mild winter areas, one-half inch to one inch will do the trick. In mixed climate zones (fairly equal heating and cooling seasons), you should use an inch to an inch and one-half of rigid foam on the outside of the wall. In areas with very cold winters, you will need to install one and one-half-inch or two inches of rigid foam board to ensure that you keep the wall cavity warm enough to be trouble free. If you check with your local building code official, they can look up what is recommended in their code books.

You can paint the inside wall with two coats of latex paint and that acts as an interior wall vapor retarder, slowing the rate of vapor diffusion. When combined with the exterior foam, this is an excellent system that works very well in any climate zone.[36] This has been called "the perfect wall" by the building science community.[37] It also works as the perfect floor when rotated ninety degrees and the perfect roof when sloped properly. It controls water vapor condensation, temperature and thermal bridging and allows drying to the inside.

In the following drawings you can see the direction of water vapor flow and therefore the direction of drying that occurs in heating- versus cooling-dominated climates. The fact that the way a wall dries is not the same in all parts of our country has led to many poor decisions and confusion about where to place the vapor

Extremely Cold Climates

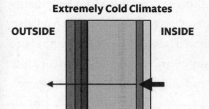

OUTSIDE   INSIDE

In a heating dominated climate zone, water vapor flow and drying take place toward the outside. This is because the Second Law dictates that heat and water vapor flow from the warmer and more humid side of the wall toward the drier and cooler side.

Layers from outside in are siding, drainage plane, wood sheathing, wall insulation, vapor barrier/retarder and drywall.

Warm/Cold/MIxed Climates

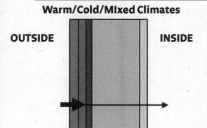

OUTSIDE   INSIDE

In cooling dominated climate zones water vapor flow and drying takes palce toward the inside of the home. This might seem unusual, but in warm climates homes actually dry in, not out. In heating climates between 1" and 2" of foam keep the wall warm enough that no condensation will occur.

Layers from outside in are siding, rigid foam acting as insulation, a drainage plane and vapor retarder, wood sheathing, wall insulation and drywall.

FIGURE 4.9. Vapor flow.

barrier in a new home. The diagrams also illustrate how the vapor barrier reduces the moisture load on the wall assembly, thus protecting it. The rule of thumb is to place the vapor barrier on the side of your wall that is more humid and warmer for the majority of the year. In hot, dry climates, walls that have no vapor barriers at all, often called breathable walls, are a good option. Also, avoid using moisture-stopping drywall products, as are commonly used around bathtubs and showers, in areas where direct water contact is not an issue.

Paybacks: Return on Investment
for Building Materials: The *Real* Cost of Housing

Remember, it's not just the initial cost of the material that represents its value. Think about how long each material will last and what it will cost to replace it over time. Also, you might talk with your insurance agent about some of the things that you could do in building your home that might lower your homeowner's insurance. Some of those things might surprise you; it would be wise to take advantage of them. Building materials that are durable, that are going to last a long time and possibly survive natural disasters like strong winds, hail and flooding reduce the insurance company's costs to repair your home. Metal roofs and stone facades are an excellent investment for this.

CHAPTER 5

Construction Waste:
The 8,000-Pound Gorilla
of Cost Savings

Money in the Dumpster

Don't you just hate paying for wasted materials? Visit a typical residential construction site and look in the dumpster. The cost of that house includes all the materials that were *not* used in its construction as well as those that were. So, all that stuff in the trash raises the cost of your building without providing you with any benefits.

Research indicates that the average residential construction project produces four pounds of waste for every square foot of living space built. For a 2,000-square-foot home, that is 4 tons of waste that usually ends up in the landfill![1] With simple efforts, that amount of waste can easily be cut by 50 percent. This means that half of the construction and demolition (C&D) waste that is sent to the landfill could be eliminated by not ever being created or by being recycled or reused. How? There are so many ways to reduce construction waste it required an entire chapter in this book!

Traditional building practices reflect a number of typical waste patterns. Conventional wood framing represents approximately 90 percent of US residential construction and uses 15 to 30 percent[2] more framing materials than are structurally needed. Construction

FIGURE 5.1. You pay for these materials—and you pay to get rid of them!
Credit: River City Rolloffs, Austin, Texas.

waste accounts for about 40 percent of all waste taken to landfills.[3] These are the same building materials that when harvested destroyed forests, wildlife habitats and other natural resources that provide vital ecosystems. Vast amounts of energy are consumed transporting building materials to local distribution centers, then to retailers and finally to the jobsite. And then to end up in the dumpster!

Many still hold onto the perspective of endless resources and the convenience of a use-what-you-need "throwaway" system, unaware of the long-term effects of these activities. More often than not, construction waste is not even on the radar for designers, builders, construction materials suppliers or site workers. The wasted materials were included in the calculation of the cost to build the home, not considered "extras" or thought of as possible cost savings.

And it's not just the wasted materials themselves that cost you money—there's also the cost of waste storage, haul off and dump fees. The cost of everything that goes into that dumpster and how many times the service company has to haul it off to the landfill and replace it onsite with another empty one to be filled again—these costs are usually viewed as "just the cost of doing business." Awareness and education in better building practices results in significant reductions in construction waste and can amount to substantial cost savings.

So, where does this waste come from and what can we do to reduce, reuse or recycle?

Reduce, Reuse, Recycle

Reduce refers to attempts made to optimize the utility and/or life expectancy of products used so that less material or less frequent replacement is required.

Reuse refers to products that have been used previously for the same purpose or another purpose and can still have a useful life. This includes products that can be removed (deconstructed) and reused again on another project.

Recycle involves taking products out of the waste stream and making them into new usable products.

Three Strategies for Waste Reduction
Strategy #1: Design for Less Material Waste

Rectangular Footprint: In Chapter 3, we talked about basic house design. You can save a tremendous amount of money on both the design of your home and its construction costs by staying as close as possible to a rectangular footprint. You will save money on footings and beams for your foundation, as these add up quickly for every deviation in the design. Then look at the impact on the structural wall and roof design for each of those projections or turns—these can also really add cost. Again, every time you complicate a design, it requires more materials to build and you create an opportunity for wasted resources. The roofer attempting to frame a corner may cut several new pieces of lumber for the angles needed.

Designing on Two-Foot Modules: Also in Chapter 3, we discussed that most building materials come in two-foot incremental lengths or widths. Even materials that are only available in four-foot width work with this strategy. If you only need three feet of plywood or OSB to finish off sheathing to the corner, cutting that off a 4×8 sheet leaves one foot of waste, or 25 percent of the material. The same is true for interior room dimensions. An extra inch or two can mean cutting into an extra plank of wood the entire depth of the room, or even worse, an extra 12-foot- or 15-foot-wide piece of roll goods. Multiply this by every place you have odd feet and inches of materials in your home and that is a lot of waste in the dumpster!

Mechanical Design: Again in Chapter 3, we discussed the advantages in designing for central core plumbing and central location of the water heating appliance for less water wasted down the drain while waiting for hot water delivery. The same is true for a central location for the furnace/air handler and a compact duct design. The added advantage of centralized design of these conventional systems is that making these systems more compact requires fewer materials that equates to fewer pieces to cut and fewer wasted remnants.

Detailed Framing Diagrams: A detailed framing plan and materials take-off lists each piece of wood, sheet and roll goods and every other material to be used in the construction of the house, as well as locations for all other items such as wiring, ducts and pipes. Detailed plans will help eliminate conflicts between trade contractors over how frame assemblies are utilized, preventing setbacks of installation schedules and cost overruns. Detailed framing plans will also help to ensure that any special structural bracing will be installed during framing and not have to be added later at higher expense. The architect should provide this level of detail on the plans. It will serve as a basis for knowing exactly what materials are required, in what quantities and how they should be installed. This knowledge can significantly reduce the amount of waste generated by contractors over-engineering structures to make up for lack of guidance.

Advanced Framing: Wood waste typically represents 40–50 percent of all jobsite waste. Advanced Framing techniques use less lumber which means less lumber waste. Framing 24 inches on center, two-stud corners, eliminating unnecessary studs for window and door framing all equate to fewer studs to cut, which means less scrap created. In-line framing minimizes the amount of waste generated from bracing and specialized framing materials that would otherwise be required to support loads. Additionally, studies on construction using two-stud corners with drywall clips have shown that this method results in fewer cracks in the drywall. This saves waste not only during construction, but also over time by reducing waste generated by repair and replacement of damaged components.

Strategy #2: Specifications That Reduce Waste

There Is *No* Substitute for Quality: Quality building materials are durable with long life expectancies and less frequent maintenance, repair and replacement cycles. Specifying quality building materials assures that the least amount of waste is sent to the landfill over the life of the home. For example, a metal roof will generally last three times as long as a composition shingle roof with less maintenance required over its lifespan. Fiber-cement siding is resistant to pests, fire and weather damage, and brick and stone claddings are durable for the life of the building, as well as reusable or recyclable at the end of that useful life. Remember to do a risk assessment and choose only materials to withstand anticipated high-risk scenarios (flood, fire, tornado, termites).

Building specifications should not only cover materials, but also the quality of work that is expected in installations. Products will only perform as well as they are installed. Specifying a durable house wrap to act as a rain screen is only as effective as the quality with which all the seams are taped, capper nails are used and penetrations are sealed. You should address quality expectations in the written scope of work provided to each trade contractor, as this is equally important for every product and material installed on your project.

Additionally, it is important that your builder check the references of all the subcontractors submitting bids for the various contracts to build the house. There is no substitute for experience and training, so make sure that all of the trades have credentials that indicate they are professionals. This is as important when it comes to proper installation procedures as it is for new products and technologies.

Remember that in some areas of the US, the only thing required to become even a general contractor is a pulse—and that requirement can be waived in the case of many subcontractors and technicians. In many areas of the country, no inspections of any kind are performed during construction. Building failures are often the result of poor craftsmanship, and most manufacturers' warranties are void if products are not properly installed. What a waste of resources to have to throw it all away and do it again because it wasn't done right the first time! It's unfortunate that we have seen this happen more than once.

Your commissioning agent should guide you in defining these specifications and to inspect the installations throughout the construction of your home. This might mean revising the construction schedule to accommodate the best series of installation practices. Performance testing is the only true way to assure you are getting the quality that you should be. The increase in performance over the life of the house will far exceed the cost of this investment.

 Turnkey Versus Labor-only Trades: Many builders prefer turnkey bids, where the subcontractor provides both materials and labor for the type of work he does. If materials are short, it's up to the trade contractor to buy more in order to get the job done, and he is not able to raise the total job price to the builder. Trade contractors typically pad their bids for material shortages in these instances, but they are also much more efficient with the materials that they do use in the construction, because it's their dime. So, when it comes to using materials efficiently, it's not likely that the framing crew will cut into a full length 2×10 for a short piece of blocking. They will have been better trained by their crew leader.

On the flip side, labor-only trades are more likely to waste beams and lumber because it's not their money and it might make their job go faster. The faster they finish, the higher wage per hour they make, since their total bid is spread over fewer hours. However, by paying crews labor only, the builder has more control over how much material is purchased, and by staging purchases and specifying waste penalties in their subcontractor agreements, significant net cost savings (and material waste reduction) can be achieved.

Reusing Materials Onsite

Brush-out Mulch for Erosion Control: Before you send anything to the landfill, stop and think about alternative uses for it. For example, you should have a site scope of work that specifies at brush-out to chip and shred any trees, shrubs and other vegetation that will be removed. The resulting mulch is great for creating perimeter berms that filter runoff for erosion control. You can do this yourself by taking a roll of 24-inch-wide chicken wire and lay it out flat on the ground. Spread a good 12-inch-deep pile down the center of the length of the wire. Roll the wire from one side to the other (wear leather gloves), binding the mulched material inside the roll to create your own mulch silt filter.

Another use for the mulched brush is to create a layer at least nine inches deep below the drip line of the trees that are to be protected during construction. Be sure to not pile mulch directly against the tree trunk, as in periods of heavy rain this can cause trunk root rot, so leave two to three inches clear space around the trunk.

Most tree roots are located in the top six to twenty-four inches of soil.[4] A tree breathes and takes in air, water and nutrients through its roots, which means the soil in the root zone area must be loose enough to allow air circulation to the roots. Unless trees are fence-protected at the drip line, any traffic across this root zone can compact soil and cut of air supply to the roots, as well as create a hardpan, which means the soil is so compacted that it will not allow water to soak in.

This slow suffocation can kill the tree over the next two or so years, long after the builder and his crews are gone, leaving you

FIGURE 5.2. Tree root zone protection should
be placed out to the drip line.

wondering why the trees, that seemed to have survived the construction trauma, would die even though you might have been vigilant in watering them along with your new landscape. Using heavy wood mulch or even a thick four-to-eight-inch-diameter bull rock layer under the drip line can protect the root zone from compaction from wheelbarrows or other construction activities. Or, if you can, fence off entire areas of the property if you have space to limit the area of the site that is disturbed and must be restored later.

 Stockpiling Native Soils for Final Grade: Topsoil, as the name implies, refers to the upper outermost layer of soil, usually the top two to eight inches. Many sites have some form of native topsoil. This soil has the highest concentration of organic matter and microorganisms and is where most of the biological soil activity occurs to support plant life.

Stop and think about the native vegetation. This soil is its preferred habitat; it survives naturally here with no human intervention. So, why would you pay someone to scrape off this soil, haul it to the landfill and then buy some inferior "builder grade" soil to finish off the landscape in order to restore vegetation? That makes no sense, but surprisingly, this is what most builders do. Sometimes the replacement soil that they purchase for final grade is utterly lifeless, contains no organic matter and has no microbial activity to support life. It's cheap, and it's a line item in most builder specifications that nobody really pays much attention to.

Most homeowners try to compensate by over-fertilizing and overwatering to try to get the installed landscape to overcome all of the shortcomings of this very poor soil. Do yourself a favor and make certain that your building specifications for site work include stockpiling the native soil for use as final grade. This will save the waste and expense of hauling it off and the expense of replacing it. Whether you intend to restore the native landscape species or have plans to do more extensive landscaping, trust that Mother Nature knew what she was doing when she put the soil there to begin with.

Use Products That Serve More Than One Purpose: Think about all of the materials and products that go into building a home. No wonder it costs so much! Using fewer products is much more effective in reducing both waste and your overall budget than just cutting the amount of waste created by the products that are to be used.

One example of this is using the concrete foundation as the finished first floor of the home. There are a number of great options for finishing concrete that complement a variety of home styles. If you are into urban modern, a sealed concrete floor can lend itself well to that industrial loft look. In a ranch-style home, it can be stained to a more rustic look. Stained concrete floors can be scored and imprinted to look like tile or even fine marble, using bordered patterns that resemble high-end custom-made flooring for each room. Polished concrete brings out the aggregate, and sealing this creates a look as beautiful as stone.

Regardless of which style you prefer, you will save having to purchase flooring materials for at least the main floor of your home. This saves on both material and packaging waste. Because concrete floors are durable for the life of your home, you will never have to replace flooring, so you save a tremendous amount of materials and more waste over time.

Natural clay plaster is another example of a product that serves more than one purpose. This material is available in a number of color choices. Selecting this finish in lieu of textured and painted interior walls and ceilings provides, again, a permanent interior finish that never needs painting. Clay helps to manage indoor humidity (supporting better indoor air quality) by absorbing excess moisture in the air.

Use Materials Made in Production Facilities

Panelized or Modular Construction: The industrial age, factory automation—historically these are credited with achieving affordability and quality in everything from automobiles to televisions. There are numerous efficiencies gained in mass production environments, some by virtue of workers skilled at specific activities, offering specialized quality control. Production work saves resources by planning the fastest process to achieve the production and defining efficiencies in the replication processes by exact calculation on the materials required with the least amount of waste. Every cut is planned, with every scrap either reused in that process or recycled to be used on the next widget coming down the line. These same principles, when applied to buildings, are referred to as panelized or modular construction, and they are the benchmark of efficiency, cost savings and waste reduction.

Panelization refers to a factory-controlled process of building wall assemblies. In other words, the exterior and interior walls of the home are constructed to architectural detail in a factory, loaded on a flat-bed tractor-trailer rig, delivered to the jobsite and hoisted into place by an overhead crane. Panelization takes a little planning, and the best efficiencies are achieved if you have numerous wall assemblies that can be built to the same production template. This means that your building designer would need to break down wall runs to standardized dimensions, with standardized window and door openings, etc. Factory panel production can include predrilling holes and running chases for wiring and plumbing pipe runs and pre-insulating wall assemblies. Structural insulated panels (SIPs) are also panelized assemblies. These can have exterior sheathing and interior drywall already hung.

Modular construction takes this concept a step further. Entire rooms or sections of a home can be constructed in the factory, hence the term modular home. Before you think "inexpensive, tract home," note that large custom homes are being offered and delivered using these same strategies. And because the waste is recycled back through the factory process, modular construction achieves the greatest reduction in materials waste onsite.

One other advantage that these methods provide is construction schedule efficiency. Since wall and roof assemblies can be in production at the same time that the foundation is happening, they can be erected onsite immediately after the concrete cures and are usually up in a couple of days (weather permitting) rather than the weeks required to stick-frame onsite. Time is money, and the cost savings from shortening your interim financing really pay off here.

Engineered Products: Engineered structural members replace solid-sawn lumber and beams, reducing old-growth lumber harvesting that creates a lot of waste in milling. In manufacturing engineered wood products, waste is cycled back through production, and in the case of trusses and structural beams, there is no jobsite waste. These products are built-to-fit exact measurements for each application, so they do not contribute anything to onsite construction waste, unlike their solid-sawn lumber counterparts. Types of engineered products were discussed in Chapter 4.

Order Only the Amount of Materials You Really Need

Remove Those Calculated "Waste Factors": Most lumber vendors will add ten percent or more to the actual quantities needed for each wood product calculation derived from your home plans. The lumber vendor doesn't want to be liable for giving a lumber quote on a project and coming up short, so they make sure that, if anything, you might have extra pieces left over. So, if the total stud count is 200 9-foot 2 × 4s, your lumber order will reflect at least 220.

Framers who know the lumber supplier has added a waste factor are more likely to cut into a brand-new 9-foot 2 × 4 when they need a 30-inch cripple stud (the short studs below windows), rather than dig through the scrap pile for a usable piece. Those kind of practices create a lot of waste.

Sometimes they add a higher percentage for waste, if your builder or the framer on the project is known for running short of materials. The same is true for the interior wood trim (millwork) order. Rather than condoning this collaboration to waste your money and waste materials, inform your lumber supplier not to add more than a ten percent waste factor to your lumber take-off calculations, or better yet, use detailed cut sheets and stage your deliveries.

Detailed Cut Sheets: Another strategy for reducing framing waste is to provide detailed cut sheets and lumber order specifications. This requires a lot of planning and detailed record keeping of where each type of lumber ordered will be used on the project. There are many ways that lumber is misdirected on projects to uses other than those intended for it. To effectively designate detailed cut sheets onsite may require counting, marking and bundling lumber for specific uses. Most contractors make notes of the various lumber size allocations anyway, and prefer to mark stacks with their allocation. This effort is meant to prevent the framing crew from randomly picking a piece intended for a specific application and cutting it for some other use, when existing scraps could have been used. Remember, the framing crew is being paid by the job, so they are glad to have the materials organized so that their time is not taken up by figuring out what materials are there or coming up short on materials, which delays their ability to complete their work. It is worth the extra effort to keep the project on schedule.

Staging Deliveries: Many green builders order their lumber in staged deliveries, with no waste factor added, starting with just the first-floor framing. With careful scrap management onsite, they can then reduce the order for the next staged delivery (second floor) by taking into account any usable remnants from the first order. Following that are the cornice, header and beam orders, and then the roof framing order. There is nothing worse for a builder than to find out that someone on the framing crew cut up a large expensive beam to frame a header over a doorway in a non-load-bearing wall. What waste! Smart builders have realized significant cost savings by managing efficiency of material use onsite.

This also works well to keep materials from "walking off the job-site." Jobsite theft adds unforeseen costs to your budget, so taking preventive measures to reduce it is important. As soon as possible, secure your jobsite by whatever means is necessary to reduce theft. If you are saving on delivery costs by having materials delivered in one shipment from a vendor that you will not be using immediately, consider renting a storage pod that can be kept secure onsite.

Strategy #3: Managing Waste Onsite

In Chapter 3, we discussed designing for deconstruction—the planned reuse of building materials at the end of the useful life of the structure. Today we see many older homes being demolished or deconstructed in order to build new, more efficient homes on infill lots in urban areas. Many items from this demolition can be recycled or repurposed, including appliances and fixtures, windows, doors, cabinetry, roofing, masonry materials and wood flooring. If you are purchasing an infill lot that was previously developed, you should make every attempt to deconstruct and salvage any buildings onsite that must be removed.

In new construction and remodeling, there are also many opportunities to recycle waste. Easily recyclable items often seen in dumpsters (actually known as "roll-offs" in the construction industry) include paper and cardboard (just about every finish item that goes into a home comes in a cardboard box), glass, aluminum and plastic (again, product wrapping, but also from worker lunch waste). These should all be sent to your neighborhood recycling facility to be repurposed into their next life cycle.

The list also includes metal (wiring, metal ductwork, copper plumbing and HVAC lines, various structural metal fasteners, rebar and, of course, nails) and clean gypsum drywall (which can be recycled to make new drywall). We can't forget asphalt shingles, which can be recycled for highway construction materials. These items may need to be taken to a specialized recycling facility that handles construction and demolition waste.

Gypsum and untreated lumber waste can also be broken up and used onsite. Gypsum drywall represents about 15 percent of total construction waste, but it can be ground and used as a soil amendment.[5] Brush and lumber can be used either for tree root protection during construction or as a final-grade landscape finish. Concrete, masonry and ceramic floor and roof tiles can be crushed or broken up onsite and used for landscape drainage features like French drains or permeable fill under pervious paving to manage stormwater runoff. Of course, any organic waste (e.g., food scraps from lunches) can be composted onsite. This is a good opportunity to go ahead and designate an area for a permanent compost site for your family to use after you occupy the house.

Signage and Posted Instructions: The importance of communication efforts for waste diversion on the site cannot be overstated. How you manage your waste on the site may vary due to constraints imposed by the area you have to work with or the availability of support services to your project location.

If you have a limited site footprint (i.e., a small infill location), you should check for a local service provider that will manage the waste recycling for you with a limited space trash bin. This allows work crews to throw building material waste into one six-foot diameter (the size can vary) wire- or wood-framed bin. The construction waste management (CWM) contractor will come out periodically, sort through everything in the bin and pull and separate it into items that can be reused in the construction, sent to recycling facility or crushed onsite for the various uses noted above.

It is important that your signage indicates which items should or should not be placed in the bin—you especially don't want food waste. It is also important to keep the waste contractor up to date on your construction schedule. If not, wood pieces that could have been reused for blocking will be wasted (mulched) if the contractor is scheduled to service the bin before blocking activities are complete.

If you think that having someone manage your construction waste is too expensive, consider all of the costs associated with traditional disposal methods. Many of these CWM contractors sometimes perform numerous other functions. Contractors have often found that by the time they cut roll-off rental, hauling fees, dump fees and jobsite cleanup fees from their budgets, the CWM bid actually reduced their overall construction budget.

If your lot is a little larger than a tight-infill size, you may decide the serviceable roll-off is your best option. Remember to let the neighbors know (again, through signage) that the container is being managed and is not an acceptable place for them to throw away their old sofa. Many roll-off companies now offer turnkey recycling management. Just like the trash bin management, the work crews throw all the construction debris into the roll-off. On their regular service schedule, the roll-off is picked up by the service contractor and taken to a recycling/sorting facility instead of being taken to the landfill. However, you should note that it is less likely that contractors will climb into a roll-off container to get scraps for reuse than with a trash bin or dedicated reuse piles. So if you are using a roll-off service, you should plan to dedicate some space onsite for a lumber scrap reuse pile.

Some of these contractors are state-licensed with a consistent achievement-of-diversion rate, so they have strictly defined limits on what can be placed in the roll-off (again, this requires signage saying what is accepted) and how much contamination is allowable. For example, River City Rolloffs in Austin, Texas, has a state-certified facility with a 97 percent diversion rate. To maintain their certification, they must have participating builders assure them that any roll-off headed to the sorting facility has no more than ten percent contamination, as sorting operations are able to handle removing a reasonable amount of contaminates, but not in excess of the stated amount. So, if excess contamination is found in any container, the entire container must be sent to the landfill, resulting in total failure of diversion attempts (not something you want to happen). This type of service is best utilized by builders who have been managing waste diversion activities on their sites for long enough to have their contracting crews well trained on acceptable management.

If you have enough space, you should consider creating onsite separation bins for managing waste. If managed correctly, this area might even be smaller than a roll-off container would require. Bins should be clearly marked (metal, aluminum/glass/plastic, gypsum, concrete/tile, cardboard/paper, food waste and, of course, other "trash"). Again, signage in all applicable languages that informs your crews of your intent to divert this waste from the landfill is a wise step. You should also attach an example of what goes into each bin using a nail gun. Nail a piece of wood to one, a piece of alumi-

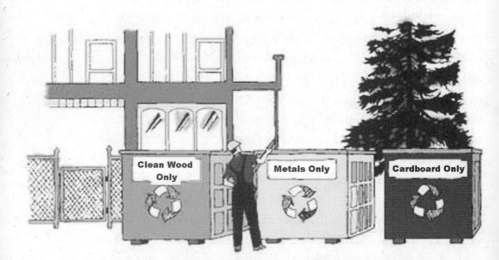

FIGURE 5.3. Dedicated waste separation bins onsite for reuse or recycling of scrap lumber, cardboard/paper, metal/aluminum. Credit: Adapted from City of Santa Monica—Office of Sustainability and the Environment.

num corner bead and a pop can to another and some cardboard and paper to another one.

Consider including terms in your vendor contractor agreements that require your framing crew to check the wood scrap pile for blocking materials, and make each contractor responsible for making sure than their crews adhere to the posted waste management procedures. This can cut your costs for hiring a CWM contractor because it reduces the sorting time. Or, you might find that you can actually take care of this yourself, saving even more money.

Safe Disposal: We also must consider how to dispose of hazardous waste products, especially building materials that contain harmful chemicals and toxins, such as lead, asbestos and chemically treated lumber and finish products (stains, varnishes, paint remover, etc.). People and companies are both guilty of unsafe disposal of all these products, including those removed during renovations. Any time we can renovate without removing these materials we reduce the amount of contaminated waste that must be dealt with. It is important that workers and homeowners are educated to help them to identify these materials. Separation and proper safe disposal methods must be used to manage these wastes to prevent environmental pollution of our soils and groundwater.

Payback: Return on Investment for Construction Waste Diversion Practices

It is impossible to quantify all the benefits from practicing waste diversion strategies. Besides saving resources and improving building performance, efficient design and materials specification strategies can reduce the base bids on key components and systems, leaving money on the table to pay for high-performance upgrades and finishes. For example, fully implemented Advanced Framing techniques can result in material savings of $500 to $1,000 and labor savings of between 3–5 percent on a 1,200–2,400 square foot house. The improvement in the thermal envelope from these efforts alone can reduce heating and cooling costs by up to 5 percent.[6]

It is certain that if we only buy what we need and manage our trade contractors so that they responsibly use every single piece of material we purchase and do not waste materials so that we have to purchase more, our costs will be further reduced. It is also certain that selecting durable, long-life materials that require less upkeep, maintenance and fewer repair and replacement cycles saves money and headaches over the life of the home. If we could add up all of the materials not wasted due to repairs over the life of the home, that is even more value!

CHAPTER 6

Equipment and Systems

Homes built today may include a number of mechanical systems, and for each function that those systems serve, a number of choices in components are available. The primary systems used in most homes are heating, ventilation and cooling (HVAC) and water heating. Some homes employ onsite sewage treatment (septic) systems. Green homes may include rainwater and graywater systems. More high-performance homes are embracing new technologies in electrical energy management, including smart lighting and appliances, as well as onsite renewable energy production systems.

Certainly advances in technologies over the last few years have resulted in far too many brands, models and variations in component configurations for this book to offer any in-depth analysis of their features, benefits or appropriateness for every application. Depending on where you live, some of these systems might not be necessary, although with recent changes in climate across the country, historic applications may not be indicative of future needs. But the main consideration for not attempting to address specifics in this book is, of course, that as soon as any words are put to the page they are out of date, as new technology and product improvements are being introduced at such a rapid rate. Having said that, this chapter's goal is to provide an overview of the general types of systems that you should investigate and consider for your project.

Spending money on high-performance systems will result in a minimal return on investment if your home is not designed for them to provide optimal performance. For a refresher on system designs, both passive and active, refer back to Chapter 3. The envelope of the home is where efficiency is rooted and comfort is to be found. Many people setting out to build a house ask, "If we have limited resources, should we focus our funds on designing and building the best envelope or build a basic envelope and pour our limited funds into the best, highest efficiency equipment we can afford?" The answer that building science studies support is clear: design, build and test the best envelope you can afford and then buy the best downsized mechanical equipment you can afford with the funds that are left. Your bills will be lower and your comfort greater in a high-performance envelope with a 13–Seasonal Energy Efficiency Ratio (SEER) three-ton air conditioner with tight ducts and the correct airflow than with a 19-SEER five-ton unit in a standard envelope with average ducts and airflow.

Efficiency Ratings

It's unfortunate that many green homes are built with expensive, high-performance systems that will never achieve their rated efficiency or deliver real comfort due to poor home or system design and sloppy installation. We cannot tell you how many homes we have seen dedicate good money to upgraded systems, only to hire an installer who is either uneducated in what it takes to achieve high-performance installations or is forced to try to install them in a home that was not designed and built for the system to operate efficiently. This is not to say that you should not invest in equipment with the highest performance rating that you can afford, it is just a caution to let you know how to prioritize your spending and set your expectations for return on investment according to your actual design conditions and the capabilities of the contractor. It is important to keep in mind that most system ratings were achieved in controlled laboratory conditions, so unless you design your home to that same standard, you should not expect that level of performance.

Equally important to mention here is the impact that the efficiency of these systems have on achieving the overall goal of net zero energy we discussed earlier. The common approach is to size your solar photovoltaic array (or other type of onsite power generation system) to cover the loads dictated by whatever your average kilowatt hour usage is. A better approach is to design and specify the home and its systems to meet the goal of reducing the total load to that capable of being managed by a reasonably sized onsite source. More on this later in Chapter 11. For this chapter, the systems are listed in the order that they might offer the greatest contributions toward that goal.

HVAC: Heating, Ventilation and Cooling Systems

In most US households, air conditioning and heating systems are the biggest users of energy, so once we have maximized our envelope efficiency, the next big target is an efficient HVAC system sized correctly to match our great envelope. This will give us the most improvement toward our overall energy reduction goals. The important considerations for heating and cooling equipment are quality of installation, correct sizing of the system and, finally, the efficiency ratings of the equipment, in that order. Before building science looked at and started recognizing system-wide equipment efficiencies, it was standard practice in the industry to oversize everything. Bigger is better right? *Not* if you want high performance, comfort and efficiency.

When equipment sizing meant installing the largest system that you could afford, the thought was that if you cool or heat your home as fast as possible, the equipment would run less and therefore use less energy. But HVAC systems, like cars, need to warm up before they reach their top efficiency. A 13-SEER system starts each cycle out at about 7 SEER and needs to run for at least ten minutes to reach its rated 13 SEER.[1] Oversized systems can cool the house so quickly that they cycle off before they even reach their rated efficiency. This is called short cycling. Boilers and other heating equipment can also suffer from short cycling and lost efficiency.[2] That means they don't have time to remove any indoor humidity. Back

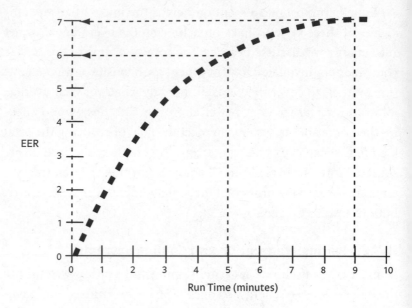

FIGURE 6.1. Run time required to achieve energy efficiency rating (EER).

to the car analogy, this is like rush-hour, start-and-stop driving. In testing conditions, the system is often allowed to run for hours so that it can achieve its most efficient operation. This is like long-distance highway driving, where our cars get their best miles per gallon efficiency.

To determine what equipment efficiency makes the most sense for your project, research the minimum efficiency ratings required by residential energy codes in your area, and then research the rating requirements for ENERGY STAR or green building programs in your area. Also, check into what rebates are available through your local utility service provider or if any federal tax credits are available. These programs will usually offer tiered incentives based on increased efficiency ratings on equipment and other criteria. Then you should model (using a building energy simulation model) the different systems that you are considering to determine which contributes best to your overall total home performance goals. The software can also estimate your utility costs for each system option and your simple payback using the various prices of the options.

Right-Sizing

The one and only way to accurately determine what size cooling and heating systems you need is by performing an Air Conditioning Contractors of America (ACCA) Manual J protocol. This is often called a heat gain/heat loss calculation or a heat load/cooling load calculation. It is a very detailed software calculation that builds a model of the home based on heating and cooling loads generated by all of the external exposed areas of the home. Each fractional square foot input is listed in terms of its orientation according to the basic eight compass directions. Every wall, floor, ceiling, window and door, along with their individual R-values and U-factors, as well as the tested airtightness of the building envelope and the heating and air ducts and any overhangs, appliances and the number of occupants in the house are included in the computer model. The program then calculates exactly how many BTUs of heating and cooling the home will require on a room-by-room basis using weather data for the last 30 years in your city. National building codes require that every newly installed heating and cooling system be sized strictly to an ACCA Manual J. If your contractor balks at this code requirement, you might want to think about finding another contractor.

This software actually includes a pie chart that breaks down the loads by building component on both the heating and cooling sides. The mechanical contractor, energy auditor or engineer creating the report will usually need to modify any default values to comply with variations that are known to exist on your project and local climate conditions. However, the devil is in the details, in that the inputs for well-sealed ducts or envelope are seldom considered, even though the software has inputs for them.

Of course, if your Manual J shows that your home is near a breakpoint in sizing, further improvements to the building envelope will result in downsizing your system size. You should note if there appear to be large loads on wall or glazing assemblies. If so, you might consider it a good investment to add some outdoor shading structures to reduce those loads. And finally, review the contractor's design for the system installation to make certain there are no obstacles in the structure that will inhibit it from delivering the premium performance that you are paying for.

Only a small percentage of residential contractors are competent in doing Manual Js, and doing them well takes experience. As with any software, the GIGO (Garbage In/Garbage Out) rule applies. You should always have the Manual J report reviewed by an independent commissioning agent, to assure that the inputs correctly reflect the climate, construction and system capacity inputs specific to your project. Or better yet, have the Manual J performed by an independent third party who is ACCA-certified to do this calculation. You can find firms online to whom you can e-mail your house plan or the CAD file and specifications and receive a cooling/heating load in a few days. You then can include it in your bid package to assure that all contractors are on the same page.

Next, the contractor should calculate the Manual S, which analyzes your load and then matches equipment selections to your home. Finally, Manual D will correctly determine the size of ductwork required to deliver the volume of air needed for comfort to each room of your home, quietly and without drafts. Ideally you want your systems sized to run constantly for extended periods of time on the hottest and coldest days and still be able to hold the set point (the temperature you select on the thermostat) with maximum efficiency and minimum wear. An oversized system that is constantly turning on and off will wear out more rapidly and never operate at its peak efficiency.

We promised you that achieving high-performance green won't cost an arm and a leg. One of the key ways to achieve this is found in the interactions between the envelope of the home and the HVAC system required. If you follow our advice to spend what is required to achieve a truly high-efficiency home envelope, here is one of the places where you will recover some of that investment.

ACCA studied sizing a few years ago and concluded that the average air conditioner in America is 150–200 percent the tonnage required by the house.[3] High-efficiency envelopes require far smaller heating and cooling units to maintain the indoor temperature. Green envelopes employing sound building science principles often need only one-third to one-half the air conditioner tonnage and heating system BTUs of traditionally built homes. This means that the air ducts are a smaller diameter and when designed correctly, each run is shorter. This means that you can budget for smaller heating and cooling units and a less expensive duct system.

Almost as important as the efficiency rating of the HVAC equipment selected is achieving the right airflow and this requires the correct Total External Static Pressure (TESP). TESP measures the resistance to airflow that your HVAC fan sees. The fan can only push so hard, and high TESP results in low airflow and reduced comfort. This happens when ducts or registers are undersized and is why we so often see rooms near our units getting enough or too much air and rooms farther away starving for air. Higher utility bills and, often, a rushing noise at the vents are other results. We measure TESP in inches of water column (wc) or Pascals (Pa) of pressure. All residential HVAC fans work most efficiently at 0.50 inches wc which equals 125 Pa of pressure. Unfortunately, field testing very often finds that they are working against one and one-half to twice that much resistance.

Your mechanical contractor should measure the resistance to airflow to ensure that it does not exceed the recommended levels, and this should be confirmed by the performance testing consultant at system start-up. This assures the correct pressure to deliver the air to the rooms in the amount needed for comfort.[4] Having enough cooling or heating capacity at the unit, but not being able to get it to each room is an exercise in futility. This situation is very common and has been found to be the cause of comfort complaints and high bills far more often than undersized equipment. Utility studies across the nation have documented the fact that half of all residential HVAC units suffer from highly inadequate airflows.[5] To be honest, HVAC equipment is almost never undersized, but it is often too starved for sufficient airflow to deliver real comfort. Don't let this happen to your new home. It's easy to measure and fix now, but very hard and expensive to do after the drywall is up.

Many HVAC contractors are afraid of right-sizing air conditioners and furnaces because they have grown up in a world of poorly built, leaky, drafty, badly insulated homes. You may have to stand your ground and insist on an HVAC system that isn't oversized and is tested and shown to have the right airflow at the right TESP. It's important to your utility costs that you do this. It is also critical if you expect your home to be comfortable. Do not let yourself be sold the old bill of goods that a bigger HVAC system than your home really needs will improve your comfort!

Another consideration for right-sizing air conditioners is humidity control. Our comfort is defined as much by the humidity level as by the temperature of the air around us. In humid summer climates, dehumidification (which is accomplished by cooling systems) requires long run cycles. This is accomplished by right-sizing the equipment, so that it runs long enough to remove the excess moisture from the air.[6] Most modern air conditioners must run for at least eight or nine minutes before they begin to lower indoor humidity levels.[7] When equipment is oversized, the resulting short-cycling does not manage humidity effectively. At higher humidity, even at a colder setting on the air conditioning thermostat, we are still uncomfortable. On the flip side, we are more comfortable at a higher temperature and lower humidity. We can actually raise our thermostat setting a couple of degrees if our air conditioner is doing a good job of managing humidity, which means long-term utility cost savings.

In humid summer climates, it is also important that the contractor select an evaporator coil and variable speed fan to maximize the system's ability to remove indoor humidity. Often these systems are controlled by a combination unit called a thermidistat. A typical thermostat only measures and controls the temperature in a home. A thermidistat includes a humidity gauge or humidistat and varies the fan speed to allow improved control of both humidity and temperature for a real comfort boost.

HVAC Efficiency Ratings

Different types of HVAC systems will carry different designation ratings for energy efficiency. Whether a heating or cooling efficiency rating is most important to you will be determined by whether you live in a predominantly hot or cold climate and based on the severity of your climate. If you live in an area that does not require much summer cooling, focus on the heating system efficiency, and vice-versa for an area with hot summers and warmer winters. If you live in a mixed climate with four true seasons, then you should be considering both heating and cooling efficiency ratings.

Cooling efficiency is expressed as the Seasonal Energy Efficiency Ratio (SEER) for traditional split systems (residential five tons and smaller) and EER, Energy Efficiency Ratio, for other types of systems. The minimum SEER rating for US manufacturers is now 13.0, and high-performance cooling is currently defined as 14.5 SEER and higher, so just getting above 15 SEER is significant. These ratings will certainly go up in years to come, so check the Internet for up-to-date guidance. Many US manufacturers have units on the market achieving 19 and 21 SEER. These units cost a great deal more than those in the 13 to 15 range. Remember, it's best from a comfort and payback perspective to put your money into a super-efficient envelope before pouring money into a very high-efficiency piece of equipment.

Heating efficiency ratings vary by the type of fuel a model uses. Gas furnaces are rated in terms of their Annual Fuel Utilization Efficiency (AFUE). The minimum efficiency for gas should be 80 percent AFUE for mild winter climates, with high-efficiency equipment rated at 90 percent AFUE or above, which mean that these units can convert 90 percent or more of the fuel that they use to heat for your home. Heat pump efficiency, for systems with capacities of five tons or less, is expressed as the Heating Season Performance Factor (HSPF). High-efficiency heat pumps are rated at 8.2 HSPF, even better ones at 9.0 HSPF. For larger units, heating efficiency is expressed as the Coefficient of Performance (COP), and a high-performance rating would be considered a 4.0 COP unit.

As mentioned previously, is important to note that these ratings are determined under controlled laboratory conditions, which are defined by the Department of Energy (DOE), in order to compare efficiencies across brands and models available on the market. Think of this like miles per gallon ratings (MPG) for your car. If the DOE did not set defined standards for measuring MPG, each auto manufacturer would use methods of testing that would optimize the efficiency of their particular equipment. "Under controlled laboratory conditions" means the units are tested for their rated efficiency over controlled run times, and, of course, the systems have been serviced to assure they are in prime operating condition, and a whole team of engineers is hovering over them like doting parents throughout the tests.

In humid summer climates, be sure that your a/c system includes evaporator and condenser coils with matched tonnages. Do not let your contactor install a mismatched system with two different tonnages. This is a common practice in the industry, but it is counterproductive to good humidity control. By matching a smaller condenser (3.0 ton) with a larger evaporator coil (3.5 or even 4.0 ton), the contractor can often claim a higher SEER efficiency. The problem is that doing this sacrifices dehumidification; the mismatched system will remove less humidity than a matched system, leaving the house uncomfortable and prone to mold growth. The physics of all of this is that the inside coil is oversized and has more surface area to cool as compared to the outside unit, so the inside coil doesn't get as cold, which results in less condensation on it and less moisture removed from the home. Contractors call it "running a warm coil." Now, if your climate zone has no summer humidity, this practice is acceptable. Again, how you build and what is right depends on where you are building.

HVAC System Selection

By the time that you have received the Manual J load model from your trusted, qualified and experienced mechanical engineer or ACCA-trained HVAC contractor, you will have already made your decisions on the most efficient building design and construction specifications for your project. Now the question is, what type of heating, ventilation and cooling systems should you invest in? You should look for higher EER or SEER for air conditioning systems, and higher AFUE and HSPF ratings for heating systems. As we mentioned earlier, depending on your climate type, usually at least one of these systems represents the biggest energy user in your home, so you want it to be the most efficient that you can afford.

The most common types of heating, cooling and ventilation systems are radiant hydronic systems and forced air systems. The more traditional forced air systems come as either packaged systems or split systems, both types consisting of the same basic components. The difference is that all components are in the same unit in a packaged system (i.e., a rooftop or through-the-wall unit), while a split system is made up of a furnace/cooling coil/air handler (blower) unit inside and a separate outdoor condenser and fan. These types of systems are available as gas, oil or electric. Gas furnaces can be fueled by natural gas or propane. The cooling components of all of these systems are electric.

Two-speed, multi-dual stage or variable speed compressors are more efficient than the old kind of blower motor because they can manage average loads while running on lower speeds, producing a colder coil and using less energy. Two-speed condensers or variable speed indoor air handler/furnace fans should be the specified blower motors in the air handlers. These are referred to as Electronically Commutated Motors (ECMs), also known as Integrated Control Modules (ICMs). They do a better job of managing humidity, especially when combined with thermidistat control units. These units achieve greater efficiency by being able to better match the load on the house under a wide range of weather conditions. Most high-rated systems include these upgrades, because they achieve their high-efficiency ratings by running like a small unit most of the time when it is not too hot, but then ramp up in capacity when it is very, very hot in the late afternoon.

In climates with long winters, the key to efficiency is in the heating system, with the air conditioner being of secondary importance. The most efficient choices are high-efficiency boilers or dual-fuel heat pumps. Here again sizing is a critical consideration. As with air conditioning, most residential heating systems are oversized by a considerable amount. This leads to short cycling, more noise and inefficient operation. Most boilers are most efficient when sized to run in longer cycles, allowing them to achieve their full rated efficiency, although the efficiency penalty for oversized heating equipment is generally smaller than for oversized cooling equipment. Other systems for consideration include hydronic heating, using pipes laid into the floors of the house, or the use of heat exchange coils in the furnace air handler. Either of these can deliver both great comfort and high-efficiency operations.

When considering a combustion-based ducted air heating system, it is of critical importance to remember that, in a tightly built envelope like yours should be, open combustion furnaces and water heaters cannot operate without the danger of backdrafting deadly, odorless, carbon monoxide; high levels of particulates and other noxious substances into your home. It is important to keep indoor air quality and safety in mind when selecting your heating equipment. Never install an open combustion appliance within the thermal envelope of a tightly built home! In tightly sealed homes like the one you will be building, only sealed combustion appliances can operate safely. This means units with 90 percent efficiency rating or higher. Even if you live in a warm climate, health and safety trump cost and efficiency in every case.

The most common types of non-gas systems are straight electric resistance systems, oil-fired boilers and heat pumps. Straight electric resistance systems include electric furnaces, wall heaters and baseboard heaters. Think wires that glow red like electric stove tops.

Electric baseboard heaters provide zoned heat controlled by thermostats located within each room. The quality varies considerably depending upon the model. Cheaper units often give poor temperature control and can be noisy. Overall electric resistance unit inefficiency, coupled with power generation and transmission losses, results in electric heat as the least efficient and most expensive method of heating your home. If you must use this type of heater, look for units carrying Underwriters Laboratories (UL) and the National Electrical Manufacturer's Association (NEMA) labels.

Although straight electric systems convert one hundred percent of the energy consumed into heat and hence they have a COP of 1.0, they pale in comparison to heat pumps, which are much more efficient. A heat pump with a COP of, say, 3.5 delivers three and one-half times more heat for each watt of electricity used than a resistance unit will. This is because heat pumps don't produce heat by burning fuel, they work by exchanging already existing heat from an external source. Moving heat from one place to another is far less costly than producing it. The most common type is an air-source heat pump, which pulls heat from the outdoor air (yes, even 40-degree Fahrenheit temperature has a great deal of heat that can be extracted) and transfers it inside to heat the home. If electricity is your only fuel choice, heat pumps are a clear winner over resistance heat and easily reduce electricity use by fifty percent.[8] In extreme cold, though, heat pumps lose their ability to extract sufficient heat from low outdoor temperatures, so these units typically have backup electric resistance heat strips or, better yet, a backup gas furnace to supplement them when needed. A hybrid unit, called a dual-fuel heat pump, combines air-source heat pump efficiency with a gas furnace backup, which greatly reduces energy consumption over the traditional electric-resistance backup heat strips. If you have a source for both types of fuel and live in a climate with a long heating season, this upgrade is something that you want to consider. These units heat the house using the compressor during milder weather, but then switch over to a high-efficiency gas furnace during spells of extremely cold weather. This means that you never have to rely on the most costly and expensive method of producing heat still in use, electric resistance heating elements.

When you get into very high-efficiency HVAC equipment, consider upgrading from the standard split system (outside compressor and inside air handler with ducted air) to a really high-COP-rated geothermal system or to an inverter mini-split ductless or multi-zoned ducted system (described below). Because they use fewer mechanical components, and because those components are sheltered from the elements, geothermal heat pumps are durable and highly reliable and typically last 20 years or more[9] with fewer maintenance requirements than most other systems. These systems are becoming more mainstream, so their technologies have reached mass production and after rebates the prices are within consideration of some traditional high-performance systems. A key consideration in the cost of a geothermal heat pump is the fact that 30 percent of the total cost of all equipment, materials and installation will be returned to the buyer by a US federal tax credit through the year 2016. Many utilities also sweeten the deal with substantial rebates based on the high EER of the units.

Geothermal heat pumps come in the form of ground-source and water-source units. These systems require the drilling of vertical or horizontal wells to pipe heat exchangers into the ground or in ponds, lakes or other waterways. With heating COPs of up to 5.0, these types of heat pumps can be one of the most efficient ways to heat and cool a home and can also provide your home's hot water needs at far less cost than any other available technology including gas or electric resistance. Geothermal heat pumps often achieve cooling efficiencies of over 32 EER. They can achieve such high efficiencies because they have an almost unlimited source to act as a heat source or a heat sink—the earth or a large body of water. The number of ground wells required depends on the size of the system and the depth of drilling that is best for your geography. Spacing on ground wells is critical, as to not overload the earth around the well components in already hot climates. Too-close spacing can build up heat, affecting system performance or even causing well failure.

Inverter and variable refrigerant mini-split and ducted multi-port mini-split systems are new to the US and Canadian markets. These are forced air systems that use a fan to push air through a coil that heats or cools it in each room. These systems move refrigerant to each zone, allowing each room of the home to be controlled independently of other rooms for both heating and cooling. Rooms that are not in use can be turned off without affecting overall system performance, and each room in use can be thermostatically controlled on its own, allowing some rooms (e.g., laundry) to be set to higher cooling temperature in order to manage higher heat loads, and other rooms (e.g., bath) to be set to heating. These are the most common HVAC systems used in Europe, Asia and South America. Although they have only been on the market a few years here in the United States, production has reached levels such that costs are within consideration of other high-performance systems, and some people predict that they may be the dominant units available in the US market in as little as ten years. Therefore, these systems are definitely worth thinking about.

If you live in a hot climate with low humidity, evaporative coolers are an economical option for staying comfortable. These units work with passive ventilation (i.e., opening the windows) to cool incoming air and exhaust warm air. They use about 25 percent of the energy and cost about 50 percent less to operate than conventional central air conditioning systems.[10]

If you are considering options that allow you some independence or backup capabilities, small combined cooling, heating and power (Micro-CHP) systems[11] produce space cooling, heating, hot water and electricity all at the same time. They work much like micro-power plants, using a turbine to make power and then using the waste heat from this process to heat the house and hot water. These systems are good in remote, off-grid locations or provide humidity control and cooling, especially as a backup power source during severe winter storms that cause electric grid failures. The electricity produced can power essential household appliances, including refrigeration and lighting, produce space heating to keep you comfortable, power absorption cooling systems or even heating swimming pools. These systems can be powered by natural gas, propane or even solar photovoltaic or other onsite power generation systems and are extremely efficient, with only about ten percent loss to exhaust, providing more than double the efficiency of electricity from the grid.

Which type of system is right for you? That depends on several factors, including fuel availability in your area, fuel costs, system efficiency and, of course, your budget (more on this is Chapter 9). There is a great calculator[12] available from the US Energy Information Administration (EIA) that allows you to input fuel costs in your area along with the efficiency rating of the various types of systems you are considering for purchase, in order to compare annual operating costs. However, when using the calculator, you should consider forecasted future costs and availability from your utility provider for each fuel type in the future. Predicted fuel supply and demand scenarios should be a primary consideration when selecting the type of system you install. The EIA also maintains future price projection reports for various fuels on their website.[13]

Installation Quality

Nothing will kill the efficiency of a system faster than a poor installation. A good mantra to remember is that "You get what you inspect, not what you expect." This was noted before, but it bears repeating here. There is no way for even a true professional to know if the airflow, total external static pressure or refrigerant charge of an HVAC system is right except by testing and measuring key parameters. If anyone tells you differently, run for the door with your hand on your wallet.

Houses are systems; think of them as being like cars. It is easy to ruin the performance of even a beautifully engineered car by making just a few errors in the fuel/air mix, timing, transmission operation or plug firing sequence. Well, a house is like that. Even if designed well, errors made in the field can easily turn a 19-SEER air conditioner into a 6 SEER or an R-21 wall into one that delivers R-10 so that your house doesn't deliver the comfort or efficiency you expected.

These errors are not made on purpose. Crews don't get up in the morning and say, "I feel like messing up someone's house today, just for grins." They are proud of what they do on the job, like most of us. But there is a shortage of skilled and well-trained people in every aspect of home construction today, and contractors across the country are finding it more and more difficult to hire knowledgeable, experienced people. That is why we should require independent third-party inspections on everything.

It is the responsibility of the HVAC contractor to provide a design layout for the duct system (according to the Manual D mentioned earlier). Their decisions here include defining whether each register will be ducted independently as a "home run" or if they will use trunk ducts. Then they have to decide which branch runs to make off which main trunk lines, where to place supply and return registers and what types of connectors to use between different duct runs. In many cases, those connectors are job built, so they must decide how to design them for good airflow.

Poor duct layout is a result of a building design that fails to take the space needs of the HVAC and its ducts into account and/or poorly trained installation contractors. It's not unusual to see duct systems that look like giant dreadlocks, with big ducts twisted and compressed through insufficient openings in floor trusses and holes cut out of structural beams (really!). It's also commonplace to see ducts run across attics and then have to make 90° or 180° turns to connect to the ceiling register, because the contractor didn't turn the register around for a straight-in shot. Poorly located chases take duct runs far off their desired path to get air where we need it, and undersized plenums create unnecessary turbulence that results in poor air distribution. These practices significantly reduce the volume and velocity of the air that will make it to its destination. These field installation errors mean that even though the duct was sized according to the Manual D software, as installed it is actually performing at only a fraction of its capacity. Also, if ducts run through areas outside of the home's thermal envelope (like in unconditioned attics), they will lose some efficiency through air leakage as well as heat gain or loss from the duct's surface.

The design provided by your architect should provide clear unobstructed duct runs with minimal changes in direction or compressions through structural members such as trusses. Like stepping on a garden hose, these compressions impede airflow, thus reducing efficient operation and comfort. The typical order of construction is for the house design to be completed and then a few months later for the HVAC contractor to walk on to the partially completed house site and figure out what might be possible. This is just plain stupid. The HVAC contractor can just as easily size and design the system based on 80 percent complete house plans and then provide the locations and sizes of the units, registers and ducts to the architect. The architect can then alter the plans to make allowances for the HVAC system components. The truss company can then build the floor and ceiling trusses with engineered spaces left open in them sized to accommodate the system and its ducts.

As was discussed in Chapter 3, placing the air handler/furnace close to the center of the house means that the supply ducts to each room or register are shorter, which also lowers the initial cost of the system because fewer materials are required to install it. It also means that the air-conditioned or heated air will have less distance to travel, so less temperature loss (or gain) is experienced through the delivery. This results in colder air being supplied to the rooms in summer and warmer air in winter. Other best practices for HVAC design were also discussed in Chapter 3. These efficiencies result in savings through lower installation and operating costs. The resulting cost savings on the initial investment should be used to offset the cost of high-performance equipment improvements.

Most of all, do not accept poorly sealed equipment, ductwork, and connections. Duct tape is not, in our opinion, an acceptable sealant. The only means of sealing duct connections for the life of the building is with mastic, and lots of it. It is not uncommon to lose 20–30 percent of conditioned air through leaks in poorly sealed ductwork. This means that you are not only losing that much efficiency in the system, but also losing that much conditioned air, meaning your return air is trying to find that much air from outside sources (that you do not want) to make up the airflow volume.

It is also necessary to seal ceiling, floor or wall boots to the structure. So, the supply boots and return boxes must be caulked to the sheetrock and floor. And, finally, check the leakage rate of the equipment itself, especially the air handler. These cabinets should be well-sealed. It is also critical that a system static pressure test is performed to assure the system is performing to the manufacturer's specifications. In new home construction, the 2012 International Energy Conservation Code (IECC) and International Residential Code (IRC) require that ducts and furnaces be tested and certified as tight and fireplaces be sealed. Homes built to this code and later versions should have only sealed combustion appliances to ensure safe operation. On existing home remodels, it is important to have an independent third party (this can be an HVAC contractor who has been approved to audit system performance by your utility rebate program) perform a combustion backdraft safety test (often called the combustion appliance zone or CAZ test). This assures that there is not a negative pressure situation, where conditions in the house might backdraft air from combustion gas exhaust piping for water heaters or furnaces. This test should be performed with all of the bath and kitchen fan exhausts turned on. Also, make sure that any indoor fireplaces have tight-fitting glass doors so that smoke and combustion gases from fireplace operations are not pulled back into the living space by negative pressure when exhaust vent appliances are operating. Airflow testing on indoor fireplaces with chimneys, but not enclosed with sealed glass doors, show they can exhaust 600–800 cubic feet per minute of conditioned (furnace heated air) out of the living space when a fire is burning.

Performing a duct leakage test should confirm that the system is sealed up enough for total leakage to be less than four CFM per 100 square feet of conditioned floor area (less than 3 CFM/100 square feet if the air handler has not yet been installed). The 2009 and 2012 IECC codes require that all new HVAC duct systems be tested using a duct leakage testing fan and certified to leak less than the code allows. The duct leakage, CFM and total external static pressure tests should all be performed before the drywall is installed. This allows for easy access to leak sites and airflow balancing dampers to adjust for comfortable airflows to each room before everything is hidden forever. Be sure that correcting problems found by these tests is a part of your contract with your HVAC contractor. It is wise to have an independent third party perform all HVAC tests. In most locales, a Residential Energy Services Network (RESNET)- or Building Performance Institute (BPI)-certified home performance technician is available to provide these services. While the total external static pressure test and the measurement of supply and return airflows in CFMs are not yet a code-mandated requirement, these tests are strongly advised.

Other performance testing is also necessary to assure you are getting a good installation. Measure the individual airflows to each room to determine that the CFM volume (cubic feet per minute) is within ± ten percent of the design volume. Remember, the Manual J software determines the exact volume of air needed to provide comfort in each individual room. This is accomplished by delivering the right amount of airflow (based on room volume) to achieve the set point on the thermostat at the same time as other rooms being serviced by the system. So the kitchen airflow is based on appliance heat loads, occupant heat loads and window heat loads, etc., while each bedroom load is based on number of occupants, size, etc. Also confirm that you have adequate pressure relief in each room by testing pressure differential between these spaces and adjacent areas to assure that these mechanisms are functioning to return the full amount of supplied CFM.

Your HVAC contractor should perform the manufacturer's recommended start-up procedures to make certain the equipment is going to perform properly. This includes: verifying the refrigerant charge by super-heat and/or sub-cooling; setting the burner to fire at nameplate input; and ensuring the air handler and fan speed settings meet the manufacturer's instructions, the total airflow is within ± ten percent of the Manual J design flow and the total external static pressure does not exceed equipment capability at rated airflow. Following these will result in more efficient delivering of the air to the room, improving comfort and saving money.

Fresh Air Ventilation Systems

If you are (and you should be!) building using tight construction methods, use your design savings on your smaller HVAC system to pay for higher-efficiency equipment and also for fresh air ventilation. Ventilation should not be an afterthought, so make certain that whatever type of HVAC system you are considering has this component included. If your area has adopted the 2012 IECC, it requires mechanical ventilation in all new homes and on any major renovation. You might think that HVAC systems (where the "V" has always stood for ventilation) include this function, but most do not. However, higher-end systems do sometimes have integrated fresh air ducting, or are integrated with separate whole-house ventilator systems.

Another method for ducting fresh air through existing HVAC systems to condition it before introducing it into the living space work by having a damper-controlled outside vent tied to a return air plenum. This simple inexpensive solution is to run a four-inch or six-inch duct from an exterior soffit to a return air plenum with a filter and damper operated by a timer that also controls the air handler. Periodically the timing device turns on the air handler, opens the damper, and *voila!* you have filtered and heated or cooled fresh air coming in through your furnace or air conditioner.

If you have more money to spend, consider an energy recovery ventilator (ERV) or a heat recovery ventilator (HRV). These units bring in fresh outside air and exhaust stale inside air at the same time. The two airstreams go through a heat exchanger where the incoming outside air is warmed or cooled while passing by the outgoing conditioned inside air. This recovers much of the energy used to warm or cool the inside air. The HRV only transfers heat between the two streams of air while the ERV transfers heat (or cold) while managing humidity to prepare the air for use in your home. This process of heat and humidity exchange reduces the impact of ventilation on heating and cooling loads. The HRV will greatly reduce the temperature difference between the fresh air and the conditioned air in the house by passing the two air streams over a heat exchanger. This will warm the incoming air in winter and cool it in summer. The ERV will do the same thing, but in addition it will pass the air through a device that will transfer much of the humidity in the outside summer air into the outgoing airstream. In this way, it reduces both the temperature difference and lowers the amount of humidity introduced into the home by ventilation in the summer. ERVs are appropriate for warmer, more humid climates, while HRVs are appropriate for cold climates.

Spot ventilation (kitchen and bath exhaust vents) should be vented to the outdoors in order to remove the moisture created while cooking or bathing. This is important for keeping indoor humidity levels in check, protecting indoor air quality, and will be discussed in more detail in Chapter 7. It is also important to note here that any large ventilation devices (exhausting 400 cubic feet per minute or more), including fireplaces and some of the larger kitchen exhaust vent models, must by code be provided with their own source of outside makeup air. There's nothing efficient about exhausting hundreds of cubic feet per minute of air that you have used good energy to condition. The loss of this air creates negative pressure that causes the house to suck in the same amount of unconditioned air from any unsealed combustion appliances, poorly sealed building assemblies or adjacent unconditioned spaces, such as garages, attics or crawlspaces.

Hot Water Systems

The second-largest consumer of energy in most homes, and the largest in homes in mild climates, is typically water heating. Any type of storage tank water heater that uses gas or grid-supplied electric power wastes a tremendous amount of power keeping water hot 24/7 while we're at work, at school, watching TV, playing outdoors or asleep. To determine which system is best for your family, you need to think about your family's hot water usage patterns. As was discussed in Chapter 3, centrally locating the water heater within the thermal envelope and having a compact plumbing design, coupled with a high-efficiency storage tank-type system can provide very efficient operations in the average home if you try to work out a schedule where you can spread out showering, laundry and cooking activities so that you can purchase the smallest-sized system you need.

The best water heating strategies use "free" energy (such as solar or geothermal). That is, it is free once you recover the investment in the system itself. Traditional water heating appliances never offer "free" water heating. These types of systems use fuels that you must purchase (electricity or gas products) over their entire useful life. They are available in numerous sizes and efficiency ratings, so if you're buying one you should purchase the highest-efficiency model that you can afford. High-efficiency gas-tanked water heaters still dominate this market. For these units, look for high EF (Energy Factor) ratings. Be sure to check the current ENERGY STAR rating criteria, as ratings are different for various sized equipment. Remember, it's the energy required to keep that water tank hot all of the time that matters.

Tankless models only use fuel when the hot water is needed, so they don't incur standby heat losses from storing hot water all the time like tanked units. Natural gas and propane are the most common. However, they usually consist of numerous elements that are fired up simultaneously to heat the volume of water demanded, so they can use a lot of fuel at once. If you have a large family, this frequent volume of energy can add up to negate any efficiency that you might have expected.

Typically, electric water heaters are not as cost efficient as gas models, since natural gas is currently the cheapest energy available. Unless you live in an area that does not have natural gas service available, and the sun does not shine enough for solar thermal, you should only consider an electric water heater as a last resort, unless you have a solar photovoltaic array and you are sizing it larger to carry the electric water heater load. This is especially true for electric tankless units, which take huge amounts of energy (often 18 kilowatts or 18,000 watts) to produce a good volume of hot water. These are only appropriate for use at remote locations that do not use very much hot water, like a handwashing sink in a workshop outbuilding. If you must use a large tanked electric water heater, make certain it has a well-insulated tank. In larger homes that require multiple water heaters of any type, efficiencies can be lost or at least significantly reduce our return on investment.

Research the quality of the product that you are considering purchasing. Look for products that are constructed of durable components with a long life expectancy that will deliver years of continued efficient performance. Also, both tank and tankless gas units require frequent maintenance, which most homeowners fail to do. This can significantly reduce both the unit efficiency and life expectancy.

"Free" water heating typically comes from either the sun or from the ground. "Solar thermal" is the term used for water heating systems that include solar panels mounted on the roof strictly for the purpose of heating water for use in the home. Geothermal heating and cooling systems (ground, air or water source) can also provide water heating. The term "desuperheater" refers to the component of a geothermal system that is used to exchange the heat it absorbs with the household water heater. During the cooling season, the heat pump takes heat out of the air in your house and, instead of dumping it outside, uses it to heat the water in your water heater. This is incredibly efficient and is close to being free. It actually makes your air conditioner more efficient, too. During the winter, the heat pump uses its ability to produce heat for half the energy of electric resistance units[14] to heat the water in your water heater at a very low cost. Neither one of these systems is literally free, but some utility rebates and tax incentives may be available to offset their initial cost. After amortizing the payback on the systems over the cost savings on the utilities necessary to provide water heating forever, the system pays for itself quickly and you do end up getting free water heating.

Solar thermal systems can also be used for hydronic space heating, such as underfloor or radiant floor heating. Or solar thermal hydronic heat exchange lines can be run through a coil at the furnace for the purpose of preheating air for space heating. Look for systems certified by the Solar Rating and Certification Corporation (SRCC) or the Florida Solar Energy Center (FSEC). These systems can be very efficient and cost effective, especially in warm or mild climates. In northern climates, however, they are much less effective in the winter, so typically they must rely on backup water heating to provide household volume requirements. Also, this type of system can be expensive, although they are also eligible for the 30 percent Federal Tax Credit through 2016.

Another viable water heating alternative may be to install a heat pump water heater (HPWH). Heat pump water heaters are air-source heat pumps that remove heat from the air to heat the water in the hot water tank. They are about three times more efficient (COP 3.0+) than electric resistance water heaters. An add-on HPWH can be used to convert an electric resistance water heater into a heat pump water heater. Be careful where you put these units, though. Since they remove heat from the surrounding air, they can serve as supplemental space cooling in the summer, but may significantly increase winter space heating loads.

Rainwater, Graywater and Onsite Sewage Systems

Most rainwater and graywater systems are regulated by local, state and sometimes federal public health and safety legislation. The concern is to prevent contamination of public water supplies from individual installations that may not be properly designed, professionally installed or maintained. For example, imagine that you have municipally treated water service to your home and you install your own rainwater irrigation (non-potable) system and tie it to the municipal water piping (to use the municipal water as a backup for periods when it hasn't rained in a while and you have exhausted your stores). If there is a fire down the road from your house and the fire truck starts pumping enough water out of the fire hydrant to cause low pressure on your supply line, without a backflow prevention valve your stored rainwater might be pulled into the public water lines. Even though you never intended for your rainwater to be used as potable water, it has now been sent out to an unknown number of your neighbors who will unknowingly drink it and may become ill. This is just one of many real world scenarios that we as responsible people must avoid. So it is important that these systems be installed by a licensed professional who complies with all safety regulations.

Rainwater systems for irrigation purposes typically consist of a gutter and downspout system installed on the edge of your roofline, and a series of underground pipes that take the captured water to one or more storage tanks. Purple plumbing pipe is used for these systems to indicate that they do not carry drinking water. It is important to have a professional rainwater system installer adjust the gutter pitch or angle in order to direct the water flow toward the downspouts, so it doesn't back up, overflow and lose precious water. It is important, as well, to provide some kind of leaf guard or at least be vigilant in keeping leaf debris build up in the gutters to a minimum, as this can seriously impede water flow and cause water losses. Some people like to install some kind of leaf guard over the downspout intakes, to remove large debris that might get washed off the roof.

If you are using rainwater for drinking (potable) water, it's also advisable to install a "first flush" tank to divert the first 50 or 100 gallons of water that carry dust, pollen and other loose water-soluble debris to a separation tank that keeps those contaminants out of your main tank. Unless your storage tank is elevated, you will need some kind of pump to get water back out of the tank and provide pressure for various uses. Typical treatment requirements for potable rainwater include a series of filters that remove sediments, contaminants and kill pathogens. Although we have heard of household chlorine treatment systems, we believe it is important that we purify our water onsite with non-chemical methods. This usually consists of a sediment filter, an activated carbon filter and ultraviolet (UV) light. For whole-house water use, typically a larger pump is required that will provide enough pressure to pull the water from the storage tank and push it through the home's plumbing. A pressure-regulating device should be installed to prevent damage to the home's plumbing system.

Using water efficiently is important to assure that the rainwater captured makes a worthwhile contribution to your needs. A graywater system improves efficiency by reusing water from tubs and shower, bathroom lavatories and laundry for flushing toilets and outdoor landscape irrigation. This system requires separate plumbing drain lines from these locations to an external distribution tank dedicated for graywater. Again, purple plumbing pipes are used to indicate that they do not carry drinking water. You should also check your local and state regulations on how long this water can be stored and how it must be treated before it can be used or discharged. Note that you cannot use graywater for edible landscapes or vegetable gardens. You also cannot use water from toilets, as this is considered blackwater, not graywater, and must be processed through some type of septic system. Blackwater often contains pathogens and disease organisms that could cause death or extreme illness. Some regulatory agencies also include wastewater from kitchen sinks or tubs and showers in the blackwater category.

Depending on where you live, if you do not have access to a community sewer line you may need to manage sewer waste onsite. Septic systems vary in complexity and cost, so check local, county and state regulations to see what type of system is required for your location. Sometimes the type of system is dictated by the amount of land you have or the type of soil, so smaller parcels or areas with shallow clay soils may require more complex and expensive systems. This is to minimize the potential for your waste to spread disease to your neighbors or the public waterways and water supply. Composting toilets may be an option; again, check for any regulations that apply. Human waste is a valuable fertilizer resource in much of the world, but it often contains pathogens that can be lethal.[15] Using composted human waste on food crops is a controversial subject due to concerns that improper treatment won't kill viruses and other pathogens, as well as concerns for pharmaceutical residues. Several US wastewater treatment facilities are using human waste to create soil amendments and fertilizers, after composting it to very high temperatures several times and then having it lab tested. Even then, they recommend that this sewer waste fertilizer not be used on gardens or crops meant for human consumption. Never use human waste by-products unless you are certain that they have been properly treated and are free from all potentially harmful pathogens.

Electrical Systems

Electrical systems are one of the most inefficient systems in a home. Usually not much thought goes in to planning how the electricity might be used, managed or protected from disruption. As the main energy-using systems (HVAC and water heating) become more efficient through advances in product technology, design and installation, the remaining "LAMEL" (lighting, appliances and miscellaneous electrical loads) represent a larger percentage of the total home load.

Only recently have product manufacturers started thinking about wasted power consumption—the power used by appliances, electronics and other equipment when not in use. Some electronic component manufacturers now include sleep modes on their systems, but only a few of those actually have built-in sensors that reduce their consumption after some period of inactivity. Other products still rely on the user to switch them to sleep mode at the end of a session, in lieu of turning the system completely off. Many home appliances still have time clocks, and many audio components never turn off, some just switching to standby mode, drawing power 24/7 and producing electromagnetic fields (see Chapter 7: Health and Environment) even after the user has supposedly turned the system off.

Peak Demand Management

There are a number of systems available in today's market that can help us to manage utilities in our homes. Many electric utility companies around the country are in the process of or have recently

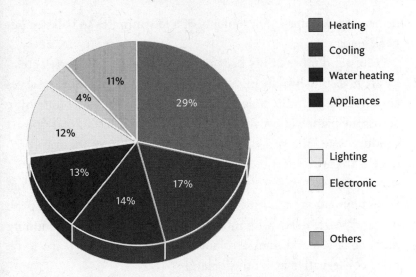

Heating
Cooling
Water heating
Appliances

Lighting
Electronic

Others

FIGURE 6.2. Electric load distribution. Credit: Adapted from Energy Star.

upgraded their electric metering equipment to gadgets called smart meters. For those of you not familiar with this technology, one of the features of smart meters is that they will allow the "utility service provider of the future" (and in many cases, "the future is now") to manage peak electrical loads by cycling off unnecessary power usage. Peak loads typically occur when everyone runs their air conditioners full blast in the heat of the summer while simultaneously doing their laundry and running the dishwasher with all the lights on. This describes the periods at which we are cumulatively demanding as much or more energy than is available at that point in time on the grid.

Whether or not this describes your family, unfortunately it does describe many others. Of critical importance is the fact that peak demand load is the basis that determines whether we must build any more nuclear, coal or other fossil fuel power generation plants. It is the peak demand load that our utility service providers must have the capacity to produce. None of us wants brownouts or blackouts. We want our power when we want it, which is a mindset that needs to change. Current applications of smart meters only give utility providers the ability to cycle off air conditioning or heating systems in increments of a few minutes to balance out peak loads.

More and more utility providers that actually produce electricity are investing heavily in renewable sources, as fossil fuel reserves continue to dwindle. Solar and wind farms dot the countryside across the planet. However, those sources have limits. Solar doesn't produce when the sun isn't shining, and wind turbines don't produce if the wind isn't blowing. And we currently have no way of storing the energy from those sources when they are producing. So, as populations continue to grow, we either meet increasing peak demand power needs with more fossil fuel generation (i.e., building more coal, gas and nuclear plants) or we as a society learn to be more conservative with our demands.

Let's say we choose conservation. It is difficult to conserve power if you do not have the ability to separate necessary and unnecessary usage. Hmm, let's define "necessary." Most people would say that keeping the power to your refrigerator on all the time is necessary to protect stored food supplies. However, keeping the power light on for the plasma television and keeping the television cable box in ready mode while you are at work and your kids are in school is probably not necessary, nor is the digital clock on the oven if there's nobody home to see it. But if these electric devices are all wired on the same circuits, how can we cut power to one of them and not others?

Electrical Load Management Systems

Structured wiring systems "home run" designated outlets and switches to a control panel box. This means that the wire goes directly to and from each of those receptacles back to the electrical panel. Even if you choose not to install any type of home energy management system initially, having your home wired in such a manner that you can control specific circuits in the future is a big plus. We'll call this type of wiring design "load management ready,"[16] or LMR for short. The first generation of these types of smart devices are now on the market, able to communicate with you and giving you the means to manage and reduce their power draw. In the future, more and more smart technology will be available for home use.

As we discussed in Chapter 3, critical load management should be the priority of the system design. As populations continue to grow, extreme weather patterns continue to occur and energy sources are more strained to keep up, having the ability to reduce total power consumption to only those systems that *must* be operable (refrigeration, water pumps, hybrid auto charging stations and critical heating and cooling functions) is what is going to be really important to us. Once we have the ability to cover our critical loads, we can then manage our other loads to reduce consumption during peak demand periods, doing our part in reducing the utility service provider's need to use those smart meters, or reducing threats of rolling brown- or blackouts, as the case may be.

Reducing Electrical Loads

Lighting and Ceiling Fans: Building codes related to energy consumption (the International Energy Conservation Code in the US and Canada) have driven huge technological advances in light fixtures and bulbs. First, we should look at the mandatory requirement in the US energy code[17] that 75 percent of all builder-installed light fixtures be high-efficiency units like compact fluorescent bulbs (CFLs) or light-emitting diodes (LEDs). Most lighting manufacturers now offer a wide array of ENERGY STAR-rated fixtures in every style and finish and for every budget that meet this requirement. One recognizable feature of these fixtures is that they do not accept a screw-in bulb of any type. This is to prevent users from replacing the energy-efficient bulb with the old inefficient, incandescent type.

Remember, electric light fixtures can give off heat, especially if you are still using the old incandescent bulbs. So, these are contributing to cooling loads, causing your air conditioner to work harder and perform less efficiently. Using daylighting windows and switching to compact fluorescent or LED bulbs and ENERGY STAR fixtures can significantly reduce cooling loads both during the day (by not having to turn on as many electric lights) and at night.

CFL and LED technologies in bulbs have also advanced to offer these products in every style and application on the market. Retrofit replacement screw-in bulbs and entire fixtures (like retrofit LED recessed can lighting) are readily available, and prices are dropping rapidly. Even high-efficacy (lumens-per–watt efficiency) light bulbs, that are fully dimmable while maintaining long life expectancy, are also readily available.

ENERGY STAR also rates ceiling fans, and these should be installed in every possible living space in the home. Air movement alone is one of the best strategies for improving your comfort. ENERGY STAR-rated fans are very efficient, and, again, any integrated lighting components are high-efficacy plug-in type bulbs only.

Lighting controls and automation systems can also save energy. These controls are available for use on individual fixtures, fixture groupings or circuits or even entire homes. Remember, the goal for energy conservation is to reduce waste, so don't let yourself be sold on a system that promotes "mood" or "scene" lighting, which in fact can result in lighting used strictly for aesthetic purposes, increasing energy consumption.

Appliances: The biggest energy users in this department are refrigerators and clothes dryers. The ENERGY STAR label has historically been the standard for energy efficiency, and many green building programs and green homes still look for this label when selecting appliances. However, their standards for efficiency have not kept up with technology, and in some product categories, it is difficult to find appliances that do *not* carry that label. ENERGY STAR is common in dishwasher and washing machine appliances, and some refrigerators/freezers/wine coolers and microwave ovens also qualify for the label. ENERGY STAR does rate any other types of kitchen appliances, (except some induction cooktop models and range hoods).

Today, the best appliances on the market are now rated by the

Consortium for Energy Efficiency (cee1.org). CEE rates appliances by tiers according to their total energy (and, as applicable, water) efficiency, currently as Tier I, II and III. The Tier III initiative, the "Super-Efficient Home Appliance Initiative" (SEHA), applies to qualified washers, refrigerators and dishwashers.

Although dryers use the most energy in the clothes cleaning department, they are not rated for energy efficiency by either ENERGY STAR or CEE. The best efficiency we can expect from a dryer, gas or electric, is to minimize its run cycle by reducing the amount of moisture that must be removed from the load. We do this by selecting a washing machine that does a good job of removing water from the clothes in the spin cycle. Ratings are based on MEF (modified energy factor) and WF (water factor). The MEF is a combination of energy factor and remaining moisture content (RMC). MEF measures energy consumption of the total laundry cycle (washing and drying), reflecting the number of cubic feet of laundry that can be washed and dried with one kilowatt-hour of electricity; the higher the number, the greater the efficiency. The water factor is the number of gallons needed for each cubic foot of laundry. A lower number indicates lower consumption and more efficient use of water.

Smart Controls: We can use smart controls on individual circuits to reduce phantom or "vampire" loads and reduce our exposure to electromagnetic fields. Plug-in smart power strips can be used to plug in various audio/visual/computer components. They work by having a master plug that controls the power to every device on the strip. For example, you plug in your home computer in the master plug socket. You then plug in your scanner, printer, fax and other office equipment associated with the computer into the rest of the plugs in the power strip. When the unit in the master plug (the computer in this example) is powered down, the strip cuts all power to the accessory devices so that they are truly off. These devices work well with TVs, cable boxes and surround sound systems that are used simultaneously. They are available either with an on/off switch for manual control or with a built-in sensor that shuts off power after some defined period of detecting no use.

If you have the ability to address your wiring from scratch, the structured wiring approach should be considered, along with some type of ability to manage each of these circuits independently. Smart electrical panel boxes are now available that can be configured to manage designated load circuits throughout your home, sensing when appliances on those circuits are not in use and shutting off all electrical power to them. Whole-house energy monitoring and management systems are still in their infancy, but many are already capable of combining smart technologies of occupancy sensor lighting, inactive circuit sensing and control and remote system controls. Phone and remote computer access allows users to adjust settings as real-life scenarios change. These technologies are advancing rapidly, so be sure to check out what components might be available as part of your overall home energy management plan on a new home as well as a remodeling project.

If your project does not include a complete wiring redo, there are some add-on smart controls that you can purchase to help reduce those unwanted loads. Whole-house outlet control systems use plug-in communicating relays that can be controlled by one switch at the entry door to shut power on and off to designated outlets. Smart light switches that act as occupancy sensors by detecting motion can be installed to replace existing switches. Entryways, basements and hallways are best controlled by occupancy sensors, so lights come on upon entering and turn off upon exiting these areas, saving energy and improving safety. Programmable lighting systems reduce electric consumption related to forgetfulness and neglect.

Onsite Renewable Energy

Onsite renewable energy generation produces electricity by methods that are privately owned and in general are meant to be consumed by the owner. Since solar and wind sources are dependent on natural daylight and wind currents, power can be used as it is generated or stored in battery chambers onsite for use during dark hours or when not enough wind is available. Some systems are off the grid, meaning that they do not have backup service provided by a utility company. Other systems are connected to the energy grid systems and have the ability to use the grid itself to bank any excess power generated and use it during off-generation cycles (i.e., net metering). The Energy Policy Act of 2005 mandates that all utility providers offer net metering to their customers. In some instances, when systems generate more power than is needed onsite, federal law, the Public Utility Regulatory Act (PURPA), requires that the utility company must purchase the excess, although at rates that can be significantly less than retail purchase rates. The Database of State Incentives for Renewable Energy (DSIRE)[18] is a comprehensive source of information on state programs promoting renewable energy, including information on financial incentives available.

Before you invest in any type of onsite power generation system, you should do your homework by researching independent third-party resources for general information on each type of generation capability in your area. Even though general information on a type of onsite generation source might or might not indicate that you are generally in an area that considers it a worthwhile investment with payback potential, your individual site conditions may vary from the norm. For example, even though your region might be known for solar photovoltaic capabilities, a particular site may have too much tree shade cover or shade from adjoining structures to be able to provide an acceptable rate of return.

On the other hand, a site on the top of a hill with exposure to direct prevailing breezes may provide enough sustained wind speeds to justify a turbine placement, even in an area not generally recognized for wind generation. An independent third party (i.e., one that is not representing a product vendor) can perform a wind study over some acceptable period of time to determine if a turbine is viable for you. You can also have that analyst perform a solar survey to determine if your roof orientation and pitch (or some other location on your property) will provide viable energy generation from a solar photovoltaic array. If you live in an area that qualifies for rebates from the electric service provider or other local

Global Irradiance Worldwide

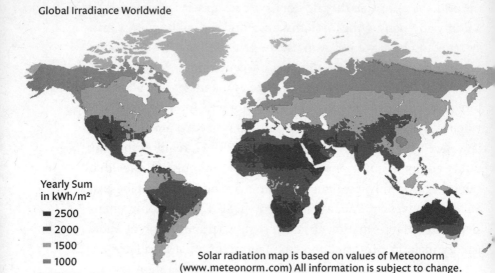

Yearly Sum
in kWh/m²

■ 2500
■ 2000
■ 1500
■ 1000
■ 500

Solar radiation map is based on values of Meteonorm
(www.meteonorm.com) All information is subject to change.

FIGURE 6.3. World solar irradiation map.

entity for these systems, they may require such an analysis for you
to qualify for the rebate. Below are the potential sources of onsite
renewable that may be worthwhile for you to research.

Solar Photovoltaics: In recent years, solar energy has gained market
share as a viable solution for providing onsite power to homes. The
most common solar energy system consists of photovoltaic (PV)
panels mounted on rooftops or ground pole assemblies. PV arrays
consist of multiple solar panels wired together in sequence, where
the total output of the array is matched to the conversion capability of an inverter, which converts the DC current generated by the
cells into AC current used in the home. In grid-connected applications, any excess electricity generated by the system is fed back
into the electric grid of the local energy service provider. During
times when the PV system cannot produce the amount of power
required in the home, usually during peak heating or cooling periods, the homeowner just purchases the additional power that they
need from the electric service provider, just as they would if they
did not have a solar system.

Generally, how quick a payback you'll get from a solar PV system depends on three things:
1. the solar access for your site
2. how much you can offset the initial investment by rebates or tax credits
3. how efficiently your home uses electricity.

The most important consideration is the last, which reminds us of the old saying, "putting lipstick on a pig." If your home is not built very efficiently for energy use, adding even a large expensive solar system may not make any worthwhile difference in your electric bill. Here we come back again to investing in your envelope before spending a bunch of money on high-dollar equipment. Before you run out to buy one of these systems to show off to your friends, we suggest you do a comparison of the amount of power it is capable of generating with the amount of power you typically purchase on your electric bill. That would be in terms of kilowatts generated and consumed. Then compare the cost of the system to the time it will take to regain that investment by offsetting the cost of the monthly kilowatts of power you may not need to purchase. Remember that solar systems only generate power when the sun is shining. Count on a few cloudy days when you are calculating how much your system will generate, as well as limiting the number of hours per day to seasonally adjusted availability. So, whatever the maximum kilowatt production is, it will vary seasonally (as days get shorter) and by how much inclement weather affects production.

Most systems do not include battery storage, unless you are off-grid. If the grid goes down, you are still probably going to be in a blackout period, along with everyone who doesn't have solar power. Battery banks are expensive and, in most cases, seldom needed. As with all types of batteries, they have a relatively short useful life. Investing in this, unless you are off the grid, is typically not recommended. Of course, research into alternative methods of storage may provide us with more reliable alternatives. As we continue to develop this technology and find ways to integrate it with other systems, new opportunities, like using our electric cars for storing excess solar generation for later use, will overcome current limitations.

Wind: The basic components of a wind turbine include a tower, turbine and a generator. The tower must stand at least 30 feet higher than surrounding structures, trees and other possible obstructions and face the direction of the prevailing wind to produce enough kinetic energy to turn the generator, which connects to an inverter and then to a distribution panel. Just like solar, any excess generation can cause your utility meter to spin backwards, creating a credit with your electric service provider. And you will use the grid to supply supplemental power when the wind isn't blowing enough to sustain your needs.

The larger the wind turbine, the more power the system can generate. For example, a 400-watt turbine can service a small number of everyday appliances, and a 5,000-watt system can generate enough energy to run most household electrical systems. Batteries are also optional, unless you are off-grid, in which case they are probably necessary components of your system.

Other Onsite Possibilities: Micro hydroelectric power will be an option only if your land happens to have a steady flowing body of water that you would be permitted to divert for that use. The flow rate must create enough pressure to rotate a turbine to create the electricity. Hydroelectric power provides 10 to 100 times more generation than either solar or wind.

Another option that comes to mind, although not very green, would be a fuel-based generator. Fuels used with these units include wood, coal, propane, gasoline or diesel. These units are typically not your sole source of power, unless you are off-grid and have no other options. However, these might be considered for a backup source of power when the primary system, usually a wind or solar power generation unit, cannot produce power due to low resource conditions. Gasoline and coal generators are expensive to operate and produce toxic emissions. Propane and wood generators can provide heat, but most do not provide electricity.

Planning for the Future: If your initial construction budget does not include funds for the immediate installation of your onsite power generation system, there are things you should consider doing if your goal is to install the system later on.

1. To start with, do the research to determine which type of onsite power generation is right for your location and site conditions.

2. Next, in planning your home design and in communicating your construction specifications, make certain that your team is aware of your future plan. In other words, if you are planning to mount solar PV panels on the roof at some point in the future, the architect needs know so he/she designs the roof with the correct orientation, pitch and flat roof area for the size of the array you will want. Also, the plumber, HVAC contractor, electrician and roofer need to know so that they do not vent any appliances in that dedicated roof space. You cannot mount PV arrays on top of roof vents. However, newer micro-inverter panels are more forgiving in placement.

3. Then think about how you will get the power from the generation source to your electrical panel box to feed power to your house and whether the power must be altered to be in a usable form. Solar photovoltaic panels generate DC current, which must be run through an inverter and converted to AC current, the usable form. With some "solar in a box" systems, each PV panel has its own micro-inverter that converts the power before it leaves the panel. Most conventional solar PV arrays, however, are configured so that the amount of power generated by the number of panels in the array is matched to the capacity of an independent inverter. That inverter has to be mounted somewhere between the panel array and the house power feed. The same is true for wind generators—they may have an integrated inverter or a separate unit. Other devices may also be required (surge protectors, disconnects, etc.), so check with a

local system installer to find out how much space you will need to mount it all and what pre-wiring you should do now to make connections later easier and less intrusive. If you need to install a metal conduit from the roof to the power feed or future solar thermal location, you might want to run this through the interior of the attic space and inside a wall cavity rather than on the exterior of the finished roof and facade. For solar thermal applications, think about where you will locate the above-average-size water storage tank and dedicate the space, along with a 110-volt outlet for that purpose (probably a ground fault circuit interrupter outlet). Also, think about reinforcing structural members to support the solar panels on the roof and having the necessary plumbing valves installed.

4. Finally, you should have your architect mark all of this on your plans, as well as note it in your building specs and Scopes of Work. You should also plan to use identification tags on any conduit, outlets or other preparations that are made for future systems, so that as time passes nobody is tempted to remove them or use them for other purposes.

If solar power is a viable alternative for your future, Natural Resources Canada's CanmetENERGY has a publication[19] on how to make your new home ready to accept solar PV generation panels or solar thermal at some future time when the economics warrant that commitment. This path is far less costly and disruptive than retrofitting these upgrades to your home later. You can download the booklet and make it a part of your building specifications. "Solar Ready" installation guidelines and other information for homebuilders are also available through the US National Renewable Energy Laboratory (NREL).[20]

Paybacks: Return on Investment from High-efficiency System Upgrades

You should expect a reasonable payback on your investment in any equipment or system. If you cannot calculate a return on investment

(ROI) within 10–15 years, it is probably not a worthwhile upgrade. Some of the new advanced technology systems, like mini-split systems, have a long life expectancy and extremely low failure rates. It is always a good idea to consider maintenance and repair costs over the life of any system when calculating what ROI to expect.

Remember, you can lower the cost of your initial investment in these systems through efficient design and integration with the other systems in the home, so that each supports the highest overall efficiency possible. The key is to integrate the system design with your architectural design by making certain there is sufficient room for placement of systems in a central location, preferably within the thermal envelope. Placing systems so that they have minimal exposure to the elements or external loads means that you can actually purchase smaller-sized systems to accommodate your needs. And remember, lifestyle choices affect both efficiency and performance. The most efficient systems in the hands of those who have no value for efficiency will return little payback on investment.

CHAPTER 7

Health and Environment

We all know that toxic substances have been introduced to the natural environment through pollution and haphazard waste disposal practices, contaminating our air, water resources and soils. Buildings also contribute to the quality of our health and our environment in a number of ways. Many of you have already become familiar with allergy problems associated with indoor air quality, including pollutants, excess moisture and stale air. But few of us think about the connection between the increasing toxicity of these environments and our long-term health.

Toxins

The Industrial Revolution enabled mass production and exponentially increased our ability to produce man-made products, as well as make improvements to existing products. While many of these developments were made possible due to the invention of chemicals, it has not been until recently, with public scrutiny and liability litigation regarding the safety of many of these substances, that there has been much activity on limiting their use. Studies by Environmental Defense Fund and the US Environmental Protection Agency (EPA) have found that the vast majority of chemicals in widespread commercial use do not have basic toxicity data published in public records.[1] Reports indicate that there are 2,863 chemicals (excluding polymers and inorganic chemicals) that the US either produces or imports in quantities of over one million pounds per year. The Toxic Substances Control Act (TSCA), passed into law in 1976, basically grandfathered the existing 62,000 chemicals in use at that time from being banned by the Environmental Protection Agency (EPA). Since that law was enacted, the EPA has only been able to require testing on a few hundred chemicals, yet the total number of chemicals has increased to over 85,000 today.[2] Of the 16,000 chemicals that have been identified as cause for concern, the study concluded that only 7 percent (just 203 chemicals) had publicly available results for all eight of the standard basic screening tests, and over 43 percent had no data in any test category.[3]

The majority of chemicals currently used in manufacturing have never been tested to determine their toxicity to humans or the environment. And under the TSCA, the EPA has mostly failed to exercise what little ability it has to restrict the use of even the most dangerous chemicals. Furthermore, the agency lacks the resources needed to provide any evaluation of chemical combinations and mixtures. The very few tests that have looked at the typical chemical combinations used today have found that the combinations are far more toxic than were their components when tested alone.

Unfortunately we are finding out too late that current research reflects that our past years of exposure to some of these substances may have contributed to the explosion of cancer, Alzheimer's, autism and other catastrophic illnesses, as well as learning disabilities and infertility. These toxins, commonly referred to as volatile organic compounds[4] (VOCs) are off-gassed by products made with those same uncontrolled or untested chemicals, including many building materials used in construction of our homes, work places and other commercial and public buildings. Breathing these toxins into our lungs can lead to respiratory issues. Many adults now suffer from allergies and chemical sensitivities. And what about our children? They spend much more time indoors since the inventions of computers, cell phones with texting and video games. This directly relates to the fact that the incidence of asthma is 3 times greater than it was just 25 years ago, now affecting 1 in 10 children.[5] Autism now affects one in 88 children,[6] a 20-fold increase in the last 20 years. Numerous research studies indicate that exposure to certain common chemical substances have been linked to these conditions.

Chemical vapors are also absorbed through our skin, making their way into our bloodstream. Ultimately, continued exposure impacts our organs, down to the cell level, compromising our overall health and leading to increased incidence of illness and weakening our immune systems' ability to fight disease. Unfortunately, often those most affected by these harmful chemicals are innocent third parties. These include indigenous people in remote areas of the world, those most closely tied to the natural environment for their food, water and shelter. Many of these areas have been devastated by mining and unsustainable forestry practices, and many of these people work in these operations with little or no protections from exposure. Additionally, these populations have little, if any, access to healthcare capable of addressing those issues.

A vast majority of all manufactured products contain some inherent risks to workers, both in factories where the products are made and those installing the products on jobsites. These workers are more likely to have frequent or continued exposures, often during critical off-gassing curing periods of these chemical compounds. In many instances, these are low-income workers who often do not have healthcare coverage, so this also represents a significant burden on publicly funded healthcare systems.

Some compounds continue to release toxic VOCs for years, though, so we are exposed to low levels but over much longer periods of time, and this exposure can cause just as devastating conditions. It is of great importance that we know how to read the Material Safety Data Sheet (MSDS) for the chemicals that will be used in the construction of our homes. By federal law, every outlet that sells a chemical must be able to provide the MSDS sheet to the public. Ask and ye shall receive. Ask your builder for the MSDS sheets for the paint, glues, adhesives, caulks, varnishes and other products that he and his subs will be using in the construction of your home. You should receive them within a few days.

Other building materials pose indirect risks to our health through their contamination of our environment, which is directly related to their life cycle processes. Human health suffers, again, from degradation of our soils, water and air, and also from the entire food chain and the delicate balance of life on our planet being disrupted. Over time, the cumulative damage ultimately threatens some species with extinction through poisoning or habitat destruction.

However, as a result of the recent increased awareness of the effects on our health of exposure to these chemicals in manufactured goods, we are seeing more development of "green chemistry." Major chemical manufacturers are now supporting a new generation of safer ingredients, recycled content and less damaging processes. Make sure their claims are supported by researching the MSDS disclosures on their products and verifying the ingredients through credible third-party sources.

Sick Buildings

Although the term "sick building syndrome" has been around for some time, mostly it has been associated with commercial office buildings, where workers sat in cubicles for eight hours each day in a highly toxic environment. Poor fresh air ventilation was compounded by years of off-gassing of commercial carpets glued down with toxic adhesives, full of accumulated pollutants and chemicals used in attempts to clean or sanitize them. Add to that the use of toxic materials in building products like asbestos (ceiling tiles, flooring and insulation), mold and mildews from past water leaks, off-gassing of toxins from laminated office furniture and commercial upholstery, high levels of electromagnetic energy, noise pollution and numerous other contributing factors, and it is not just the buildings that get sick. The syndrome itself refers to the higher incidence of workplace absenteeism, reduced productivity and higher insurance claims.

But how many of us ever dreamed that we would find similar situations in our homes? The very place that was supposed to be our safe haven may turn out to be the main source of our toxic exposure. The average American spends approximately 90 percent of their time indoors, where air quality can be 2 to 5 times (or sometimes up to 100 times) worse than outdoor air quality.[7] In other words, we have a high exposure to these unseen contaminants. Sometimes odors or smells give us a clue, but our reaction has been to cover these up with air fresheners (more chemicals!) or scented candles (petroleum-based wax toxins plus more chemicals for fragrance!). How many of you reading this still think that that "new car smell" is a good thing?

What Are These Toxins and Where Do We Find Them?

The most important materials that affect indoor air quality are those within the walls of the structure, including interior finish products. Most interior wall construction is air and vapor permeable, so toxins within the walls can seep out, impacting indoor air quality. When insulation, subfloor materials, synthetic gypsum and even glu-lam or other engineered wood beams off-gas toxins from the binders used in their manufacturing processes, these toxins readily enter your living environment.

The National Toxicology Program of the US Department of Health and Human Services publishes a report[8] that identifies substances and exposure circumstances that are "known or reasonably anticipated to cause cancer in humans." A couple of the chemicals covered in the report are commonly found in building materials.[9] Formaldehyde, specifically urea-formaldehyde, is used in the binders of many common building products, including plywood, particleboard and fiberglass insulation, and styrene (used in the manufacture of some carpets, carpet backings and carpet adhesives). Currently polystyrene-based foams dominate both the spray foam and the rigid foam board insulation markets. Additionally, the flame retardant used on spray foam products also off-gasses toxic VOCs.

These products are a combination of chemicals that must be mixed in proper ratios and applied under appropriate climate conditions by appropriately trained professionals for both performance and safety. Unfortunately, many times the crews doing the mixing and installation have not been properly trained. Sometime the weather doesn't comply with their hurried work schedules. Chemicals not properly mixed and installed release toxic VOCs at much greater rates.

Even properly installed spray foam has at least a 12-hour curing period. Most of the toxic off-gassing will occur during this period, so it is best to cross the full day after the foam installation off your construction schedule and not allow any other work inside the house during that time. By manufacturer specifications, no one is to be inside the home during spray foam installation unless they have the required protective clothing and dual canister respirators. It's best to open all the windows and allow the building to air out for the day. It is especially important that drywall not be hung during this period, as drywall is an air barrier and will trap off-gassed VOCs within the wall cavities.

Strategies for Improved Indoor Air Quality

We should first practice *source control*. The idea is to avoid introducing contaminants into your home in the first place. Any efforts here will significantly reduce both your indoor air quality issues and the efforts and the expense that you would otherwise incur attempting to remediate them. This sounds obvious, but on how many projects does the builder spend even ten minutes considering these issues? The sad answer is very few, unless you make it a point to discuss this issue and insist it be a consideration. We now understand that our homes, like our commercial buildings, must be properly ventilated to control the buildup of toxic chemical compounds and particulates. When we ventilate, we dilute these potentially damaging contaminants and restore a healthy environment in our homes. Building science research has a mantra you may want to remember: "Build the home tight, but ventilate it right."

Many product labels contain cautionary instructions for wearing safety apparatus while applying the product and proper ventilation of the space, but do not give you any indication of the time frame of off-gassing that might affect the health of your family. To minimize our exposure to harmful VOCs, we need to become more educated on the different types of common products used in home construction, as well as those used to maintain the home over its lifetime. VOCs can be reduced or eliminated from interior finish products by specifying products that are tested and certified to be within acceptable limits.

You should also look for interior finish products that contain no heavy metals, phthalates or aromatic solvents. It would be wise to reference a known and respected source, like Green Seal, with the professional staff needed to evaluate MSDS sheets and provide you with acceptable standards specifications for various types of materials, the guidance that you will need to make the right choices. Even better are products made from natural ingredients with less environmental burden associated with their life cycle, that are durable, with ease of maintenance, and do not require toxic cleaners.

Materials Finished Offsite

Whenever possible, choose products like cabinetry and flooring that are prefinished before they are delivered to your jobsite, or have been finished offsite. This allows the initial blast of high-VOC off-gassing to occur before the product is ever introduced into your living space, where continued release of toxic fumes will be less. This, of course, is not true if the product has then been wrapped in plastic to keep its moisture content stable or to protect that new finish. If this is the case, it's best to remove the plastic wrapping and leave the materials in an open-air garage or storage space for a day or two before installing them. Of course, the amount of off-gassing varies, depending on the type of product, so some products will still release VOCs for years even if they came prefinished, like engineered wood floors or cabinetry. Again, you can request MSDS reports on these products, as they are required to provide information on the chemicals used in these processes.

Materials That Require Less Finishing

As mentioned in Chapter 4, selecting materials that require less finishing can save resources and reduce maintenance, thereby reducing costs both initially and over time. However, perhaps the most important contribution of less finishing is less off-gassing of toxins associated with finish products. Natural wood and stone products, as well as tile and concrete, can be used without toxic finish applications or with natural finishes that are durable and have a long life expectancy.

Radon

The Environmental Protection Agency (EPA) recognizes that "radon is a cancer-causing, radioactive gas...that is estimated to cause thousands of deaths each year...the second leading cause of lung cancer." If your home is to be located in an area recognized as a radon hazard zone, you should take the appropriate steps to properly vent and reduce the risk of exposure to occupants within the structure (see Chapter 4). Other odorless soil gases, such as methane, may also be a concern and should be addressed through appropriate detection devices.

Garages

Many homes have attached garages that contain bags of chemical fertilizers and pesticides, volatile gas fuels stored in poorly sealed containers, automotive exhaust fumes, paints and varnishes and other toxic products. Every time we open the door from the garage to the house, we allow some of these fumes in. Even with the door closed, the wall separating the house and garage is typically two and a half times leakier than other exterior building walls. Why? Homebuilders don't see this as an outside wall, so they don't take the time to effectively air seal it. Make certain this detail is addressed if you are building or remodeling this area of your home.

We recommend moving all chemicals, fuels, fertilizers and pesticides from your attached garage into a detached garden shed. And never start your car with the garage door closed. Installing an exhaust fan in the garage that is tied to the garage door opener and runs for a few minutes after the car has exited and the door has closed is also a good idea. The US EPA has made some recommendations about how this could be achieved at their website.[10]

Laundry Rooms

Laundry rooms are major sources of heat and moisture (breeding dust mites, mildew and molds). They should be equipped with an exhaust fan, vented to the outdoors. These fans should have a sensor to activate operation automatically and run until the sensor detects that the desired level of moisture reduction has been achieved. Clothes dryers contribute to negative air pressure in the home, exhausting large volumes of air that must be made up from somewhere, which usually means from places that you would prefer not to draw air from. The best case would be to locate the laundry area outside of the conditioned space. Installing a clothes line will also contribute to reducing these laundry room issues.

Combustion Appliances

Gas combustion appliances, including furnaces, water heaters, fireplaces and cooktops release toxic gasses (carbon monoxide, carbon dioxide, sulfur dioxide) every time they are ignited. Standard, open-combustion appliances are not sealed well enough to prevent backdrafting from the combustion gases in the vent pipe in the event that negative pressure occurs in the home. This can happen anytime the clothes dryer, fireplace or exhaust fans are in operation. In a tightly sealed home always buy sealed combustion gas appliances and always use proper exhaust spot ventilation when operating gas cooktops.

Fireplaces

Indoor gas, pellet and wood-burning appliances and fireplaces also release toxic fumes into the living space. EPA researchers suggest that exposure to wood-burning appliance emissions presents a greater cancer risk than exposure to the same amount of cigarette smoke.[11] Additionally, the fine particulates in the smoke emitted from chimneys is the largest single source of air pollutants for neighboring homes.[12]

The worst offender is the ventless gas-fired fireplace, which have been outlawed in a number of states and countries. Even though these appliances claim to be 100 percent efficient, as their name implies they have no means of exhausting to the outside the combustion gases and particulate pollutants generated by their operation. All of those gases and particulates are sent directly into the living space. Additionally, these appliances generate a lot of moisture, which can further impact indoor air quality, leading to mold, mildew and related health concerns.[13] You should never install an unvented fireplace inside your home, due to the high risk of contaminating your air or causing damage to your family's health or even death. In fact, it is better not to have any type of fireplace or wood-burning stove installed indoors on a porch or patio, unless it is EPA-certified as meeting required safety standards. Direct-vent fireplaces work the best to manage smoke and combustion gases and take them out of harm's way.

It is also important that fireplace appliances have a dedicated source for outside combustion air, so that they are not pulling conditioned air out of the living space to create the fire and then exhausting that air up the chimney. If you have an existing fireplace and would like to make it as safe as possible, make sure it has an air vent and then install sealed glass doors, so it can be operated with the doors closed.

Fresh Air Systems

Introducing supplemental fresh air ventilation to increase air exchanges also improves indoor air quality by diluting the level of contaminants caused by off-gassing, soil gas infiltration or infiltration of toxic fumes from all of the sources that we have discussed. Because high-performance building systems provide typically tighter construction than traditionally built homes, they do not allow the recommended quantity of fresh air to infiltrate the envelope. Mechanical fresh air systems not only provide a calculated amount of fresh air into the home, but also should provide some filtration of the air through the conditioned air system, so allergens and pollutants introduced to the space are minimized.

Filtration

Air filtering is also a good front-line defense for protecting the mechanical equipment from contamination, although not so much for improving the quality of indoor air. Although many filters manufactured today remove some percentage of dust, pollen, dust mites and other contaminants, the truth is that they do not manage the turnover of air within your entire space well enough to keep the air particle-free. Some return air grilles are located on walls near the floor, while others may be located high on walls or on ceilings. Some are in areas of the home far away from the main sources of pollutants (like in hallways, far from exterior doors, moisture sources or garages). Many spores and pollen grains fall quickly to the carpet only to be stirred up every time you walk through the room. To think that filters can be effective in cleaning our air is not realistic.

High-efficiency particulate air (HEPA) filters meet the standards of efficiency set by the Department of Energy (DOE) by removing at least 99.97 percent of all particles greater than 0.3 microns from the air that passes through filtrations systems.[14] The key to improving indoor air quality is to have these systems ducted so that they are able to continuously circulate the large volume of air within the conditioned space, which is not an easy task. Additionally, these filters only remove airborne particulates, so they are not effective in removing heavier particulate matter that settles out, or at cleaning the air of toxic fumes off-gassed by indoor pollutants. Return air grille location high on walls or on ceilings is very effective at reducing particulate matter contamination of ductwork, since gravity will keep most of that below the intake zone. It is important to note that filtration does nothing to reduce the volume of toxic vapors caused by the off-gassing of chemicals in building materials. The best means of addressing good indoor air quality is to exchange fresh air with stale air (see "Ventilation Systems," Chapter 6).

But it is a good idea to use good filtration and do the best that you can to protect your system from premature contamination problems, and also to do what you can to remove dust and pollutants from the air. At minimum, pleated media filters should be used at all return air grilles. Filters with a rating of MERV 8 fit this basic requirement. MERV 10 is better and MERV 13 is best, as long as your HVAC system fan is rated for the restricted airflow of these filters caused by their design. Many residential furnace/air conditioner fans are not powerful enough to pull air through the thicker or more efficient filters without suffering premature failure. IF you intend to use a MERV 10 or higher filter, you should inform your HVAC contractor so they can determine if the fan in the system they are installing is compatible. Change filters at least according to the equipment manufacturer's recommendation or more frequently if your family's lifestyle and health concerns require it.

Keeping Water Out of Your Structure

Water rots almost all building materials, causes mold to grow and ruins the ability of most insulations to function. Great green framing lumber, surrounded by green insulation will only serve to grow a great crop of mold and will decay into rubble if water is allowed to intrude. In Chapter 4, we provided details on how to flash windows, doors, porches, chimneys and other key areas of the exterior walls and roof to ensure your home remains a place of dreams, not nightmares.

The basements and crawlspaces of homes are commonly places we avoid because of their musty odors, mold and dampness. A clue for you: when you smell a musty odor, what you smell is mold. As with spoiled meat, our noses are acutely tuned to that smell through thousands of years of exposure and self-preservation, and we very rarely mistake it for any other odor. Building science has proved through a million blower door tests that the basement, attic, garage and crawlspace air are intimately connected to our living spaces. Some people try to tell themselves that those bad things are down there and not in here. If only that were true; but it isn't. The air in these connected areas moves into your home easily through HVAC ducts that run through these areas, the plumbing and electrical penetrations that connect them, hollow interior wall cavities, utility chases and floor framing. Even if you don't have a mold allergy, your wooden framing is going to suffer structural rot and decay if it is exposed to high moisture levels.

Humidity Control

Humidity control should be the first defense against indoor contaminants from mold, mildew and dust mites. At relative humidity levels over 50 percent, dust mite populations flourish, leaving unseen masses of dust mite waste that creates havoc for humans sensitive to that respiratory contaminant.[15] Mold grows on surfaces above 60 percent relative humidity during the cooling season and in building cavities where humid air collides with cooler surfaces and condenses.[16] All of these activities may go unseen but not undetected by sensitive noses. And again, our lungs are the doorway to our bloodstreams. Many HVAC systems now come with humidity-sensing thermostats (thermidistats) that cause the air handler to operate when unacceptable humidity levels are detected. Air conditioners function as dehumidifiers, so running them helps to bring humidity back to tolerable and healthy levels. Best practice is to have whole-house humidity control systems analysis recommend systems designed specifically to address these concerns.

In some very cold regions, humidity control includes the use of a whole-house humidifier. If you live in an area with long periods of very low relative humidity, these devices can add to your comfort. But a word of caution is needed here. When indoor relative humidity levels reach 30 percent or above during the heating season, the danger of condensation forming on cold surfaces increases greatly. This condensation can occur around windows, on basement walls and other cool surfaces. The resulting mold growth and possible decay can be very harmful to your family's health and to the durability of your home. If you find even slight dampness forming in cool areas, lower the indoor humidity level until the condensation stops. It is also very important to remember to turn off your humidifier when warmer and more humid spring weather returns.

Moisture Control and Spot Ventilation

Spot ventilation can help to remove heat at the source, lowering the load on your air conditioning system. Use of bath exhausts, laundry room exhaust and kitchen cooktop exhaust also help to manage moisture at its more noticeable sources, helping to keep indoor moisture and humidity levels in check. Unfortunately, most people do not understand the effective use of these appliances.

Oversized kitchen cooktop vents (400 CFM or larger) create negative pressure that can literally make a house suck in pollutants. The makeup air will follow the path of least resistance and that often means from the least desirable sources. How many of you have turned on the cooktop vent only to have the fireplace not draft properly, filling your house with smoke? This is because the fireplace offers the closest source for makeup air for that large volume of conditioned air leaving your home every minute through the exhaust fan vent pipe. If you don't have a fireplace or if it is sealed with tight-fitting glass doors, the next path of least resistance is probably any unsealed gas appliance vents. If your home has unsealed gas appliances, your best solution is to seal up any closets housing the appliances with weatherstripping and ensure that they have access to combustion air from outside.

Many bathroom exhaust vents are not used correctly. Although most can exhaust 50–100 cubic feet per minute of source-point moisture created during showering or bathing (if they are properly vented to the outdoors), most people do not run them long enough to remove that moisture. Or, you might do the opposite, turn it on and then forget that it is running. These appliances should be run 20–60 minutes (depending on the length of their duct run) after each bathing event to effective do their job.

Today we have some great new options for bathroom ventilation. First, there is a new generation of bathroom vent fan. It has a permanent magnet motor so it runs on far less electricity and runs far more quietly, too. In fact these new vent fans make less than ten percent of the noise of the fans of just a few years ago and are therefore virtually silent. The controls available today also make the proper use of these fans almost foolproof. There are now wall switches available that contain relative humidity or condensing sensors. They can be programmed by simply turning a set screw to come on whenever the humidity in the room exceeds 50 percent or condensation occurs. Others have occupancy sensors that turn the fan on when the bathroom is occupied and off when you leave (of course there may be times that you visit the bathroom without needing to operate the fan). If you specify the use of only ENERGY STAR-certified exhaust fans[17] that meet ASHRAE 62.2 standards, you can ensure that your new fans will be both quiet and energy efficient.

Interior Finish Products

Many adhesives used to glue wood together, install floors and countertops and bind product components together have formaldehyde and other chemical compounds in them. So ask questions about the products *and* the materials that will be used to install them. There are formaldehyde-free and low-VOC options available for all of these products, and many carry no additional cost.

Cabinets, Countertops and Trim: Many cabinet materials are made using urea-formaldehyde binders, as are subfloors, decking under countertops and some interior trim products. Most cabinet vendors purchase the components from production sources, especially the boxes, drawer assemblies and doors, which can be made from MDF (medium density fiberboard), particle board, or OSB (oriented strand board that may contain these binders. These products release toxic VOCs into the indoor air that your family will be breathing for years.

Alternative binders with much less harmful ingredients can be used. MDF that is certified to meet certain standards is available for both cabinetry and millwork. Look for cabinetry carrying either the CARB (California Air Resource Board) or E-1 (European Standard) certification.

Flooring: The key considerations for flooring products are that they be hard surfaces. Stay away from carpet and sheet vinyl, even those products certified as green off-gas toxic chemicals for years, and carpets trap pollutants that have been tracked into homes and shed from our clothing and shoes. Some of those pollutants are organic, meaning that they are a food source for organisms higher up on the food chain and therefore attract these pests into the home. All of these organisms create waste products as they go through their entire life cycle—breeding, pooping and dying—in the carpet or other crevices of your home. Carpet is virtually impossible to keep clean.

Choose materials that are prefinished or finished offsite with low-VOC finishes and are installed onsite using minimum adhesives (i.e., floating floors). Engineered wood floors (not solid wood, but wood flooring manufactured through a layering process) may also use formaldehyde binders and glues. So, again, check for products made with non-urea-formaldehyde binders and finishes, choose low-VOC adhesives, stain the concrete foundation or use natural solid wood, tile, linoleum, cork or bamboo.

Finishes/Adhesives: Paints and stains, varnishes, solvents, binders, caulks and sealants are big sources of VOCs. VOC content is usually listed on packaging as grams per liter. Safe limits vary per product type and are defined by the Green Seal GS-11 and GS-47[18] and other reputable standards. There are many products available on the market that are within the recommended VOC levels published by Green Seal, some at little or no additional cost.

When you think about sources of VOC off-gassing in materials, do not forget about the floor adhesives, floor finishes, countertop adhesives and wood trim finishes. Also, think about how the product fits into your long-range needs. Moisture-cured urethane finishes are durable but too toxic. You might find penetrating oils not durable enough or requiring too frequent maintenance. Contemporary waterborne finishes can provide the same durability as tough oil-based products with much lower environmental consequences.

Other Health Concerns
Electromagnetic Fields

Electromagnetic fields (EMFs) exist everywhere in our environment. These fields can be created by both natural sources, like thunderstorms, or can be man-made, like x-rays and high-frequency radio waves, with sources including radios, TVs and cell phones. EMFs are produced by electrical power generation, transmission and use, including all of our home's wiring and all of our electrical systems, fixtures and appliances. EMFs at different frequencies interact with the body in different ways, depending on their frequency or corresponding wavelength. Since chemical reactions within our own bodies produce electrical currents, exposure to certain EMF fields can interfere with or cause a physical reaction, including changes in cell structure, tissues and hormones. How our bodies react to EMFs can vary due to our overall constitution and health. Research linking EMF exposure and a variety of health concerns is ongoing and, for the most part, inconclusive. There have been some fairly consistent patterns in EMF exposure and incidence of certain leukemia and other cancers. Of course, some individuals may be more sensitive to EMFs than others, and continued or high levels of exposure may have a higher impact on their health.

However, many, including the World Health Organization and the European Commission, agree that limiting our exposure is just common sense. This usually means distancing yourself from sources and limiting exposure times and frequency. To prevent any problems arising from continuous exposure to EMFs in the home, choose a building site that is not close to large overhead distribution transmission lines for electric utility services. Electric fields from power lines rapidly become weaker with distance and can be greatly reduced by the walls and roofs of buildings.

Your electrical design plan should allow for controlling circuits to limit your exposure, especially in sleeping areas, as our bodies are much more susceptible to these fields when relaxed. Avoid locating main living and sleeping areas in EMF hot spots,[19] such as near electrical panel boxes, main service house connections, etc. Also, avoid running wiring to major electric appliances overhead or through walls of sleeping areas. Devices are available that can be programmed to shut off these circuit when you turn off the table lamp next to your bed. Since many clock radios, clocks on microwave ovens, electronic clocks, computer and television sleep modes mean that these appliances are actually in some state of "on" 24/7, having a switch to cut power to the circuits or a smart control that recognizes when they are not being used can significantly reduce our overall exposure.

Environmental Contaminants

Manufacturing processes related to building materials, from mining to production factories and transportation of products, often result in air, water and soil pollution. With oversight provided by the Environmental Protection Agency (EPA), the pollution generated from manufacturing processes is monitored and measured to legislated "acceptable" limits, although some might argue that any contamination is not acceptable. To reduce pollution, we must develop less polluting methods for extracting materials and manufacturing building materials. Products manufactured from recycled or waste materials, or from rapidly renewable resources have much lower impact on the environment than those sourced from mining and harvesting of virgin resources. Products made from locally sourced materials or manufactured locally have much lower embodied energies than those sourced and shipped long distances. It also generally uses less energy to ship raw materials some distance to a local factory than to ship finished goods over great distances. Bulk shipments of wood components or rail cars full of metal ore have a lower environmental impact than the number of shipments required for the volumes of finished manufacturer goods, especially when each container holds only one finished cabinet box or kitchen appliance.

Sustainable Manufacturing Practices: Many product manufacturers have improved their processes to produce less waste in manufacturing. Additionally, some manufacturers have voluntarily adopted and implemented more sustainable manufacturing practices, both as cost-saving measures and in order to be better citizens. The International Organization of Standardization (ISO) has a voluntary Environmental Management Standard program that supports manufacturers in setting and monitoring goals aimed at improving their environmental stewardship. ISO14001[20] is the third-party designation for companies that participate in that program.

Greenhouse Gases: Air pollution in the form of ozone, particulate matter and greenhouse gases in our atmosphere can all contribute to climate change. Burning fossil fuels to generate electricity in order to power our homes, industries and automobiles, as well as the methane gases created from large cattle operations and landfills, have exponentially increased the amount of greenhouse gas released and trapped in the atmosphere. Building construction activities contribute to these issues. Strategies to reduce the carbon footprint related to our housing and lifestyle choices are discussed in depth in Chapter 14.

Light Pollution: Another type of pollution that is not commonly recognized is that produced by outdoor lighting in the form of light pollution, also known as light trespass. Many nocturnal creatures are dependent upon dark skies to hunt for food, find water sources or navigate their annual migration patterns. Light pollution interferes with their normal functions and can adversely affect their ability to move around, choose safe nesting places or avoid their predators. One example of this is reflected in a study[21] that found baby sea turtles that naturally use the direction of star and moon-light reflected off the water surface to help them find the ocean when they emerge from their beach nests may turn the wrong way and migrate toward the brighter lights of buildings or streetlamps, often resulting in death.

Contributions to light pollution come from our outdoor porch lights that throw light up into the night sky, soffit floodlights that light up long distances in our yards, billboards, business signs, street lights and parking lot floodlight systems and numerous other sources that we no longer take notice of. Almost everyone has seen satellite images of our planet at night and the huge areas that are blasting light out into the cosmos. Of course, it takes electrical power to generate all of this light, again contributing to more greenhouse gases and climate change.

Few of us ever consider the impact that this 24/7 light has on our health or that of other inhabitants of the planet. Plant life is dependent on daily and seasonal natural light patterns to produce fruit, shed and grow leaves, grow roots and stems and go dormant. Light pollution affects these natural cycles and can result in permanent evolutionary changes to some species.[22]

Therefore, it is important that we take steps to minimize light pollution and trespass wherever possible. If your home will be located where nighttime light pollution is unavoidable, it's wise to take steps to ensure that your bedroom windows have the necessary treatments to create a truly dark sleep environment. For safety lighting, choose exterior light fixtures that are shielded from above, styles that have a solid top that prevents light from shining upward. Also, limit landscape lighting and select fixtures that throw light exactly where you need it, without lighting up unnecessary areas. Choose fixtures specifically for path or deck lighting and landscape fixtures that spot particular features in the landscape. Better yet, limit outdoor lighting to what is needed for safety and security. If you feel the need to use floodlights, make them motion activated and set sensitivity to respond only to larger predators (like a burglar), not small, nocturnal creatures.

Noise and Other Pollution: The hazardous effects of noise depend on its intensity, pitch, frequency and duration. Exposure to over eight hours of sound in excess of 85 decibels a day[23] is potentially dangerous. In addition, noise can be harmful to animals and the environment. Noise can disturb normal functions of wildlife, including feeding and breeding.

Sources of noise pollution that affect the home include vehicular traffic and aircraft, commercial and industrial activities, HVAC system operations and common household appliances, like vacuum cleaners, exhaust vent fans and lawn equipment. Methods to control this include reducing exposure, wearing ear protection or using something to block the path of transmission. Site selection for our homes should include an analysis of exposure to noise-producing pollutants, like proximity to highway traffic or industrial manufacturing. Trees and shrubs can also disrupt sound waves from external sources, as does insulating the exterior walls and ceilings of our homes. We can also sound-insulate the interior walls of internal sources like media, laundry or bathrooms. There are new types of drywall that have very high sound transmission class (STC) ratings, which measure how well a building partition attenuates sound. In the case of a media room, sound-attenuating drywall, double layers of drywall and offset double stud walls, are all good ways to control interior sound sources. Internal sources of sound pollution should also be considered and dealt with during the design phase, such as locating the HVAC air handler in the attic or in an insulated closet.

The same concern could be expressed for pollutants that affect our other senses, including smell, taste, vision and touch. We would not want a home located downwind of a landfill or close to a stockyard. We would also find it difficult to live on a landscape reclaimed from a toxic chemical dump, so best to think about these things when selecting your building site.

CHAPTER 8

Outdoor Living

The cost of building a home goes beyond the walls. The cost of acquiring and developing the land it sits on can be considerable, so every dime spent on this should contribute to the overall quality of life in the home, not just be an overhead expense. If we buy building sites in planned developments that include shared amenities such as parks, playgrounds, swimming pools and recreational facilities, then our monthly homeowner association dues are an economical way of enjoying these types of amenities without each household having to bear the full cost or duplicate those resources. Regardless of what your community offers, there are benefits to using your own individual homesite:

- an extension of the living space.
- our personal connection to the natural environment.

An important part of green building is its association with nature, and this is important not only to protect the environment but also for our physical and mental health. Just about every area in North America has some season conducive to spending time outdoors. Breathing fresh air and the visual interaction with nature lifts our spirits and puts us in touch with wildlife and the fact that we are not alone on this planet. It connects us with the planet's state of health and increases our awareness of our human contribution to its delicate balance that keeps us healthy as well.

Every home and community should value its open, outdoor space as much as we value what we build. To make the best use of your land, you must extend the scope of your design beyond the building structure. Be sure to include outdoor amenities that will add to your quality of life.

Defining Your Outdoor Living Spaces

In Chapter 3 we discussed the considerations important to defining your outdoor living space as part of your initial design of your home. You should designate your outdoor space in a place where you would most likely want to spend some time during mild climate months. You're not going to use a space that is difficult to access. Covered porches might actually be part of the home's building structure, or they might be independent structures, like decks with pergolas. Even without a porch, a seating area in the landscape can serve the same purpose. Investing in some outdoor furniture will provide you with a standing invitation to use it.

Many times this outdoor living space is added after you move into the home, over time as your budget allows. It is important to develop a long-term design and stick with it. Think about and plan your plantings around these spaces even if they don't exist yet, so that you are not redoing work later. Plant deciduous trees on the east and west sides to provide shade for these spaces in the summer yet allow solar heat gain in the winter when they drop their leaves. Carefully space all trees and plants to accommodate the mature size they will become. Most plant labels provide spacing or full-growth size information. Consider how groupings of plants will look in designing the space based on mature size, leaf size and shape, bloom color and season, whether the plant will die back in the winter or stay evergreen, compatible watering needs with companion plantings, light exposure and shading requirements and how much care will be required to keep the landscape design enjoyable.

Patios and Porches

Whether you are planning open patios or covered porches, it is important to think about how these fit into the overall design of your outdoor space and how you will use them over time. For an outdoor living or dining space, a pergola or other shading device may be all that you need to use that space through most of the year. This might even be a deck with a shade arbor for growing a lovely wisteria vine on for deciduous shade. Flatwork patios have more limited use as your family and friends may not find them that inviting in the heat of the summer.

Many people won't spend time outdoors due to concerns about pests, especially mosquitoes, flies, wasps, spiders and other creepy-crawly insects. Screened porches are an excellent way to overcome that obstacle. Any screened outdoor living space will allow you to enjoy the spring, summer and fall seasons without being bothered by mosquitoes and other insects. All of these outdoor areas are much less expensive to build than conditioned space.

Outdoor Fireplace

Is there a place you could put a free-standing outdoor fireplace, dig a fire pit or build a portable or permanent firebox? Or maybe some seating near the fire pit that would allow you to enjoy star gazing? Since we tend to reduce the amount of time that we spend outdoors during the winter, adding an outdoor fireplace or fire pit can also improve that. This experience is close to camping in our own back yards, again increasing our awareness of what is happening with nature, the planet and wildlife.

Summer Kitchen

Especially in hot climates, cooking indoors in the summer adds unnecessary heat (and moisture) to the living space. This really increases the loads on air conditioning equipment, increasing utility cost and the burning of fossil fuels to stay comfortable. Outdoor summer kitchens keep those problems out of the house. At minimum, plan a space for an outdoor grill. If possible, integrate this area with other outdoor living space, including a fireplace and dining area, so you are getting both the benefits of being outside a greater portion of the year and reducing your need to keep your indoor thermostat setting as you would if you were indoors.

Outdoor Shower

For those who enjoy gardening, having an outdoor shower to rinse off garden soil and keep cool while doing summer chores can be a big bonus. However, an outdoor shower can also serve as your main shower in the milder seasons and summer months and is a great way to keep moisture out of the house. Anytime you can reduce indoor moisture, you are protecting air quality in the home. Runoff from the shower is considered graywater, so you should check local codes to determine if you can use it in your non-edible landscape as "free" irrigation water.

Landscaping

Now it's time to think about the greenery. The outdoor landscape provides you with your own slice of nature, even in the most urban areas that don't feel natural at all. You should decide what area you want to embrace for outdoor living and focus most of your landscape planting efforts there. Why waste your money landscaping areas of your yard that you will never spend time in? These areas just become a chore to maintain and water, and thus cost money not only to install, but also for their upkeep. We often wonder why homeowner associations mandate minimum landscape requirements for "street appeal" when other non-living scapes could be as attractive, require less maintenance and use little or no water.

Once you have defined the areas that you will enjoy as landscape, look out to the rest of your yard and beyond. If "beyond" leads to green belts or similar natural areas, think about how you will transition your landscape areas to blend in with those natural areas.

Native plants and grasses make beautiful landscapes. Nobody wants to spend time in or maintain a landscape that is not appealing to the eye. Use mulch or rock beds to reduce weeds and natural organics like corn gluten to stop weed seed germination. Space plants so that they won't overcrowd when they are fully grown and mix colors and blooming frequency for seasonal interest. Creating your own little oasis will support biodiversity and establish a self-sustaining environment where your family can enjoy the many health benefits nature provides.

Soil

The most successful landscapes start with good soil. If you have clay or sandy soil that lacks good drainage properties, adding soil amendments like organic compost will greatly improve its ability to hold water while still allowing air to the roots of plants. Adding compost is as important as any other strategy for long-term irrigation cost savings and a thriving landscape. The money that you spend to improve the quality of your soil before doing your landscape will pay you back many times over the years in reduced need for supplemental irrigation and dealing with other problems that can result from stressed plantings.

Integrated Pest Management

You can minimize the risk associated with termites by using a couple of integrated pest management strategies while designing and installing your landscape. These include planting with the main plant structures or trunks at least 36 inches from the foundation, as mature plants planted too close to the exterior walls can create a bridge for termites. This also keeps the foundation visible, so annual walk-around inspections will leave termite tunnels from the ground up the wall visible so that you can address the problem before too much damage has been done. This is one of the reasons that your finished grade should leave 8 to 12 inches of exposed foundation visible as well (in addition to reducing risk from major stormwater runoff events).

Plant Selection

The most important landscaping advice is to select plants that are native to your area and appropriate for the location you are planting, with the right sun or shade requirements. In areas with less rainfall or subject to drought conditions, you should limit plant selections to native or drought-tolerant plants that require a minimal amount of watering. If you live in the desert, cactus and succulents are good choices, as well as native grasses or grass species that do not require frequent watering, as they can take the heat. On the other hand, if you live along the coast with high humidity and frequent rainfall, you need plant selections that will not drown, get root rot and are not susceptible to fungus or mildew-related disease. Areas that have high rainfall (over 36 inches per year) should consider landscaping trees and plants that do well in boggy soils.

Check your local agricultural extension office or local water department for a list of recommended plants for your area. These native species have proven their ability to survive in your climate with little or no help from man once they become established in your yard.

Think low maintenance, low water use and no need to use pesticides or chemical fertilizers. This means continued money savings, as these activities add periodic costs even after the initial installation and stabilization period. However, we will offer one note of warning: if you do not provide deep watering during the plant's establishment period or if you continue to frequently water a plant after it has rooted in the landscape, you will make it dependent on that watering method. Even plants that would otherwise develop deep roots or be able to go long periods between watering will become slaves to an irrigation system schedule. In the event that irrigation schedule ends abruptly (because of water rationing during drought cycles, for example), even those native plants will not survive.

Turf grass needs more water than most landscaping species. For long-term water conservation and water-bill savings, limit turf areas. If you need a green space for your kids to run free in, decide where to put that space. Otherwise, lose the turf areas; they only contribute to a higher water bill. Usually a well-mulched bed of native butterfly- and bird-friendly plants will require less water per square foot than a lawn and add more interest and color to your yard. Choose ground covers that don't need mowing or watering. Or put in a rock garden, rain garden or mulch bed.

Check references to make certain you are not introducing an invasive plant to your landscape. Just because you can buy a pretty plant at your local nursery (and it's a native) does not mean it is a good addition to your yard. Invasive plants will move outside your defined plant beds into your lawn, seed pods can blow in the wind and take over your neighborhood before you know it, sometimes choking out other good plantings.

Trees

Trees along the east and west walls can shade these from direct sun exposure. If you live in a predominantly hot climate, possibly even in an area with little or no winter, having evergreen trees on the east, and especially on the south and west sides of the house might be the best strategy, especially if these provide shade for some of the roof area during the heat of the summer months. Deciduous trees will lose their leaves in the winter, so they also provide seasonal passive solar shading, protecting the building from heat in the summer and allowing heat gains in the winter, regardless of what climate zone you are in. Plan your building site to take advantage of these natural elements if they exist or, if not and you have space, you should plan to plant some well-adapted, fast-growing tree species, with at least one-and-a-half-inch-diameter trunks when you finish construction. The best time to plant trees is always in the fall, although we've heard an old saying that "the best time to plant a tree is yesterday." Place trees in your landscape more for the shade that they will produce for your structure and outdoor living spaces than for aesthetics.

If your site is located in a dense urban area, trees contribute to reducing the phenomenon known as urban heat island effect. This situation occurs in urban areas due to the large expanses of concrete and other paving absorbing heat from the sun during the day and then radiating that heat back out into the immediate area.

This can typically cause these urban areas to record temperatures around ten degrees warmer than surrounding rural areas and can really be a problem during hot summers, especially for low-income and elderly residents who cannot afford air conditioning. It can reduce your ability to enjoy your outdoor living space. It can also cause problems for outdoor pets, as well as birds, insects and other wildlife.

Erosion Control and Maximizing Natural Irrigation

You or your site management consultant should develop a permanent erosion and stormwater management plan to take advantage of the natural drainage patterns on the land so that they continue to provide the most cost-effective natural irrigation possible. If possible, gear up with your rain suit and umbrella and walk around your property during a heavy rain, when the rain is falling faster than it can be absorbed into the ground. Look at water flow patterns. This might be easier on a sloped site, but even on a fairly level site, the water has to flow somewhere.

Years ago, the prevailing strategy was to get the water off of the lot as quickly as possible. In urban areas, we installed a stormwater system that had intakes at locations along city blocks and underground piping to take the water to local creeks and remote disbursement areas. Unfortunately, this resulted in major erosion of native soils, as well as washing away chemical fertilizers and pesticides into our lakes, rivers and creeks. Runoff from our yards also picks up and carries with it pollutants from automobiles left on driveways and roadways. Add to this the chemicals from lawn equipment, and you've got a toxic cocktail that contaminates our surface water systems. This pollution creates algae blooms, poisoning aquatic life and our drinking water supplies and impacting downstream farmlands and fish. This is especially true when we disturb the ground during construction, removing the native vegetation that holds the soil in place. As a result, that native vegetation will not be naturally restored once the disruption is complete if the native soils are no longer present.

So now we treat residential construction sites basically the same as we do commercial development: we manage stormwater through a variety of strategies that are intended to keep it onsite to be used as natural irrigation. You should have removed any native topsoil and stockpiled it before construction began on the site. Now is the time to restore that soil prior to landscaping with native plants. This is also the time to develop permanent stormwater management features to channel the water onsite where it can infiltrate slowly, providing deep natural irrigation to your landscape.

Sometimes it may be necessary to install a French drain to capture water that might threaten the home's basement or foundation and get that water quickly to another area. This is a great way to use leftover concrete waste, or rock that has been unearthed during the site preparation process. Using materials that are otherwise considered waste not only keeps these materials out of the landfill, but saves you money that would have been spent hauling it off and also spent purchasing other materials to use to create the drainage field.

By incorporating berms, swales and/or rain gardens to direct and capture rainwater, we can reduce native soil erosion and keep our stormwater local to replenish our groundwater supplies, while using the earth's layers to naturally cleanse and filter it. A rain garden is an excellent way to reduce erosion and capture valuable rainwater, providing it more time to infiltrate. This reduces our need for supplemental irrigation of the landscape and, at the same time, replenishes our groundwater supplies. It's easy to install —it's like deliberately creating your own low spot in your yard. Just dig an impression into the ground, any size that fits the space you are working in. Use the soil that you remove to build a berm in the direction that water naturally flows across your yard, so that you are directing the water into the low spot. Plant landscape species that can tolerate having their root zone wet for a few days after each storm, but these same species should be able to tolerate dry spells as well (because they are developing deep root systems). If you search the Internet for "rain garden plants," you should be able to find species appropriate for your area and climate.

Wildlife Habitats

What if your site is not adjacent to a natural area? Be creative! Don't you wish it were? Why not create your own wildlife park? Look into dedicating space for wildlife: a butterfly garden, bird sanctuary, squirrel feeder, etc. Do hummingbirds appeal to you? Certain species of plants provide the right combination of color and nectar sweetness to attract them to areas where you might not have seen them previously. The same is true for beautiful species of butterflies and birds. The truth is that, when you landscape with native plants, you will attract the native creatures that depend on those plants for survival. Many beneficial bugs lay eggs, feed on and find shelter in these plants. You may find that you enjoy the wildlife as much as the wildflowers. Check with your local plant nurseries or extension service for native plants that attract these wonderful creatures to your garden.

Even if you don't become an enthusiast, keep in mind that beneficial insects keep the balance of nature intact. This means using organic methods and products for fertilizing and pest control. Using chemicals to green your lawns or control weeds also kills beneficial insects and microorganisms, and may poison other forms of wildlife that are grazing on or nesting in the landscape.[1] Never use a weed-and-feed product. Atrazine, their active ingredient, has been found in the drinking water supplies of most cities, and it is known to cause a host of health impacts in people.[2] The same is true for the herbicide (weed killer) known as Glyphosate. If you are unsure as to which products contain these compounds, read the labels, check their MSDS, or Google the chemical name to find products that contain it.

Check out having your yard certified as a wildlife habitat. For more information, check with the National Wildlife Federation; it offers federal certification for "Certified Wildlife Habitat."[3] Many states and municipalities also offer similar certifications. It actually doesn't take much to accomplish this beyond what you would be doing for a native plant landscape. Just make certain your landscape plant selections provide habitat (shelter and a place for wildlife to raise their young), food (nectar or seeds) and a constant water source (bird bath, pond). Note that these features also add to your views and living enjoyment.

If a wildlife habitat is not part of your design, at least consider implementing a no harm policy in your plan. Light pollution from exterior light fixtures, landscape lighting and floodlights affects the feeding, mating habits and migration patterns of numerous nocturnal creatures and birds. Be sure to store any chemicals out of reach of wildlife and make certain no chemically contaminated water (with fertilizers or pesticides) is left accessible. Choose to use only natural organic fertilizers and pesticides in your landscape. Make an effort to eradicate any invasive plant species and not introduce any non-native invasive plants to your landscape. If possible, you might also consider whether you can dedicate at least part of your landscape to nature, making it inaccessible to pets or other animals that might be predators.

Rainwater Collection

Add rainwater collection containers to roof gutters to provide supplemental water to landscape plants when needed, as well as for wildlife support. Using treated water supplies for landscaping is a waste of the energy used in the potable water treatment process. And your plants will love you for their rainwater! Be creative with your rainwater collection capture by incorporating berms and swales to direct water toward your planting beds regardless of whether or not you invest in water storage tanks.

CHAPTER 9

Green Bling

The first ten chapters of this book are intended as a roadmap to building a healthy, high-performance home on *any* budget. So far, we have walked you through all of the cost-saving aspects of location, size, design and resource efficiency in Chapters 1, 2, 3 and 5. In Chapters 4 and 6, we presented the important considerations for building materials and methods, as well as equipment and systems. In Chapter 7, we discussed every aspect of building a healthy home. Finally, in Chapter 8, we recognized the benefits of incorporating outdoor space to extend our living area and protect the environment.

The goal that this book offers is to make building or remodeling green homes something everyone can achieve. These chapters have not been just a series of steps for you to follow. They are also a priority list of what will give you the biggest improvements in performance and cost savings. The key to achieving *this goal* is in the early steps. If you fail to follow those recommendations, you will just be adding expensive upgrades to an inefficient project and may never see the return on those investments that you expect.

Green bling refers to expensive green upgrades. These range from high-performance heating and cooling equipment, high-efficiency water heaters and plumbing fixtures, energy-efficient appliances, photovoltaic electricity panels and home automation systems, to solid wood and stone floors, earth-plaster walls and custom cabinets. It is unfortunate that many certified green homes earn that designation by adding these expensive upgraded systems without making the effort to first improve the building performance so that they realize a worthwhile return on those investments.

In green building certification circles, we call this "chasing points." In most programs, higher green building certification levels are achieved by scoring more points from à la carte lists of recognized green features. Too often program participants plan their project by selecting items that will give them the most points, without regard to how these individual items work synergistically with all the other features of their home or whether they provide a worthwhile return on their investment based on the specific conditions of the project.

Let's look at a few examples of this add-on point strategy. First, there are those who go for the highest point-value categories, like adding a solar photovoltaic system, but fail to make the energy efficiency improvements of Advanced Framing, shading devices or improved thermal and air barrier systems. Yes, they rack up the points, but that expensive system quickly erodes their budget and offers little in terms of return on investment because the house itself will not operate efficiently.

The same is true for rainwater collection systems to be used for outdoor irrigation if you live in an area with frequent and sufficient rainfall. These systems eat up a large portion of your budget yet would provide little benefit. So it is important to consider the return on each investment we make based on their value in meeting our needs. If there is no need, there is no value.

Unfortunately, what we see too often is that projects start off with the list of bling that they want in their home because they believe this is what defines a green home. They most often fail to take any measures to address the inefficiencies in the base costs that would provide funding for these upgrades. In many cases, homebuilders are handed a set of off-the-shelf home plans and a list of the desired green bling features that they then have to figure out how to achieve within a specified budget. They are often forced to cut costs some way to make these fit into the budget. That leads to the use of inferior-quality materials, such as cheap batt insulation, and low bids that result in substandard installations in order to cover the costs of high-efficiency HVAC systems and tankless water heaters. These projects usually end up over construction budget and never achieve significant operational savings because these system upgrades were not matched to a well-designed, high-performance building envelope.

Seldom do these methods of adding green bling ever deliver the intended performance. And this approach to green building has led to the misconception that green building is considerably more expensive than conventional building. This approach is also the fastest way to blow your budget and end up with a home that does not provide the performance that you expect. This is not what this book recommends.

As we stated at the beginning of this book, there are many strategies and methods that, if utilized, can result in green buildings that are not more expensive to build and will save tremendous resources, including your money, over time through their efficient operations. Some of these strategies rely on using trade-offs to rebalance the home budget so they must be implemented throughout the process.

The Modest Green Home

But, again, we promised you that you can build a green home on *any* budget. We make this claim even knowing that some of you are on very limited budgets. If you are in this group, we suggest that you focus on achieving every aspect covered in Chapters 1, 2 and 3, as well as the basic recommendations that follow from the remainder of Section One. You will want to limit the amount of cost savings achieved that you spend on any green bling in order to stay within your budget. It is important, therefore that we provide guidance for you to understand which of the strategies discussed both fit your budget and provide you with the greatest benefits. This focus will guide you to sustainable, passive strategies in order to build a resource-efficient home, and doing so will provide you with great benefits over traditional building practices. So before we look at the bling, we want to recap the steps that we took to reduce our baseline budget and improve our home's performance so that we get the greatest efficiency out of the home's mechanical systems.

First, and highest on our priority list, *we designed a home that was resource-efficient in every way.* This means it saved materials resources in construction, and it will save energy and water resources over its life cycle. Even if you are building in the suburbs, following the strategies in Chapters 2 and 3 will accomplish the passive design benefits and related cost savings. If you do build in the urban core, following the recommendations in Chapter 1, you will also save our community resources in managing the infrastructure and services needed to support the home over its lifetime, thereby helping to keep all of our property taxes affordable, as well as your commute time and expenses, and giving you access to alternative transportation choices, both of which protect our communities outdoor air quality.

Even with the most modest budget, you should still use quality building materials and methods, to assure the long-term durability of your home. This will reduce repairs and maintenance that can adversely affect your ongoing household budget. It is important to review the "How We Use Materials" section of Chapter 4, to assure that your wall assemblies provide the best water management and thermal performance. Even if you cannot afford more expensive insulation or rainscreen materials, how you use what you can afford is most important. Advanced framing is key to reducing costs and improving performance.

As much as you can, you should include products and systems to improve indoor air quality and basic system-efficient operations. Usually this includes spot ventilation, low-VOC wall and ceiling paint, hard surface (or no) flooring and right-sizing your HVAC coupled with a good-quality installation. In smaller homes, mini-split systems may provide the best investment for conditioning limited spaces. The more you can do from the following list, the closer you will be to achieving your goal.

- The location is either urban infill or within a mixed-use community with existing infrastructure and in a walkable neighborhood with bicycle and mass transit access.
- Home size is limited, accommodating only our defined long-term needs, with spaces that provide flexible use for our changing needs over time or common needs for future owners.
- The design is site specific based on a simple basic rectangle with total wall lengths on two-foot increments.
- The site development protects or restores native habitats and captures stormwater for natural irrigation.
- Turn-key contractors provide expert input into the design to achieve high-performance assemblies and operations.
- The roof design is a simple gable or hip roof, with a passive solar long axis ridge line running east to west, with unobstructed roof orientation to the south.
- The structural design of the building envelope permits a continuous and contiguous thermal enclosure.
- The plumbing design is compact with the water heater centrally located between wet locations.
- The plumbing design incorporates water reuse to the extent possible by code in the area.
- HVAC furnace/air handler are centrally located with short runs to inside walls.
- Large overhangs and porches protect walls and windows and provide shading appropriate to the climate to reduce cooling and heating loads.
- Windows provide daylighting and passive ventilation.
- The room layout minimizes internal heat loads and optimizes comfort working with passive ventilation strategies.
- Durable materials, resistant to termites or other pests, are installed with methods that reduce the risks inherent to your location.
- Wall assembly materials and construction methods are used that withstand water damage, restrict airflows, provide high thermal value and manage water intrusion.

- The HVAC system is sized according to Manuals J, D and S with correct inputs for climate, design and construction quality, with a quality installation.
- The design provides basic fresh air ventilation.
- Spot ventilation is provided at cooking appliances, laundry rooms, and bathrooms exhaust moisture outside of the home.
- Electrical design provides the future ability to manage phantom loads.
- Lighting design provides options to provide specific lighting needs and controls.
- Appliances are sealed combustion (furnace and water heater, if they are gas).
- Uses low-VOC interior paint and hard surface (or no) flooring.
- All contractors are provided written Scopes of Work detailing the expected quality of installation and resource- (cost-) efficient strategies.
- Strategies put in place to minimize waste while optimizing performance.

These strategies will lower your base construction cost, help to keep your total cost of living low over the life of living in the home and significantly improve the efficiency and comfort of the home. You will have achieved a very affordable and comfortable home. Even if you have very limited budgetary constraints, you can still use what money you have to build a home that provides passive benefits and reduces the burden on the community as a whole.

The Mainstream Green Home

However, many of you will plan to reinvest the money that you saved to achieve a deeper shade of a green home, and yet still may be limited to the median average range on how much you are able to invest. If this represents your scenario, our recommendation is that you first follow the same advice and do the basics recommended above. In fact, these are the same strategies that you must use if you are going to accomplish a worthwhile return on investment on any level of upgraded materials and systems if you expect real high performance and healthy benefits.

Beyond the basics, our next priority is to focus on more durable, high-performance building materials, methods and systems and improvements to indoor air quality (Chapters 4, 6 and 7). These materials and methods continue to improve our building performance, lowering our long-term total costs of ownership. Once you have a high-performance envelope, it's likely that your next best investment will be from a more efficient HVAC system, if this is the biggest energy user in your home based on your climate. Focusing on these upgrades will enable you to reduce your energy loads so that you will save even more money because you can install a smaller HVAC system.

So our first level of green bling reflects the types of products, methods and systems that we often see in production building developments. Many times these homes are speculative offerings that builders develop based on targeting a specific market and price point. The features in all of the homes within a community will be the roughly the same, including the base building material specifications and system selections. Homebuyers may have choices in interior finish out, street elevation design and exterior façade materials. This allows buyers a level of customization, but the basic bones of these homes stay consistent so that the builder benefits from volume purchasing discounts with trade contractors and vendors. The level of green bling included in these homes often includes materials and construction methods that are based on risk assessment for our site and proven building science for our climate that will provide a durable, long-life building envelope. These criteria assure the homebuilders that they will have happy customers that will provide repeat and referral business to them.

Some of these strategies will provide additional cost savings, so we want to reinvest those savings for long-term benefits. Production builders set their budget limits based on the market price that their research has shown will attract the desired buyers. Every dollar added to the budget must be offset with some cost savings in order to stay within the defined market. These markets are often competitive, though, so builders do try to negotiate every upgrade possible in order to sell out as quickly as possible, as their interim financing costs can significantly affect their profitability. This tier of green bling is aimed at what would be reasonable for builders and buyers in those markets.

All green homes should also start with the basics of location, size, design, durable materials and construction methods. Developments in the urban core can benefit from marketing shared amenities, including recreation and green space, and close proximity to community resources, including commercial and retail establishments. Design, as was mentioned in Chapter 3, can do more than anything else to reduce base construction costs to allow these builders to add more competitive green bling features. The competitive edge should start with the biggest impact items. Improve your thermal envelope according to Department Of Energy ENERGYSTAR recommendations for your climate.[1] Once you have a high-performance envelope, it's likely that your next best investment will be from a more efficient HVAC system, if this is the biggest energy user in your home based on your climate. Have an energy model prove each component's contribution to improving performance. Also, keep an eye toward building materials and finishes that support healthy home features and lower maintenance requirements. The following items will give us the biggest bang for the buck in this level of high-performance home:

- Improved wall and roof assembly materials and construction methods based on sound building science (e.g., adding exterior foam insulation)
- All ductwork and mechanical equipment within the thermal envelope
- Durable exterior materials with long life expectancy, including exterior façade, trim and roofing
- Durable interior finishes (solid-surface countertops, hard surface flooring)
- Materials specified that are locally harvested or manufactured (marketed as supporting the local economy)

- Materials, where possible, that serve more than one purpose and are panelized/modular or engineered for resource efficiency
- Higher efficiency air conditioning and heating systems, if this is a major energy-using appliance for your climate, with programmable thermostats
- Mechanical fresh air ventilation
- Outdoor living space
- Building science and energy modeling to assess the best building and mechanical system components to optimize total building performance
- Performance testing to locate any remaining leaks and confirm an effective sealing of envelope and ductwork

Using resources efficiently means that we continue to provide more funding for high-performance building assembly materials and methods that will offer years of low utility and low maintenance cost savings, and supporting infrastructure savings for lower property taxes. Energy modeling can confirm which materials and methods provide the greatest utility cost savings. Building a tight house is also the best thing that we can do for indoor air quality if you upgrade to sealed combustion appliances and include fresh air ventilation. Performance testing assures us that we have achieved the desired quality of installations for those benefits.

The Custom Green Home

For those of you who have the funds available for building or remodeling a custom home with high-performance systems and upgraded finishes, you'll need guidance on which will give you the performance you expect. It is important that you discuss these with your commissioning agent and have an energy modeling consultant provide feedback as to which have the greatest impact in your climate. Even with a higher budget, we must continue to look at green bling in terms of return on investment, where your added dollars pay back in as short a time frame as possible. The only way to accomplish this is by continuing to assess the impact that each investment has on our building-as-a-system's overall performance.

In custom homes, we have a greater selection of high-performance systems, so again, which one you select will depend on your budget. Depending on whether you are in a cooling- or heating-dominated climate, look at higher EER/SEER-rated air conditioning or higher AFUE/COP heating systems. Two-speed or variable-speed compressors and air handlers offer far better performance than single-stage systems. Thermidistats or independent humidity control devices are also worthy components. The best (and more expensive) options are ground-source heat pumps, multi-port (or whole-house) multi-split, or micro-CHP systems. In very arid climates, it is worthwhile looking at evaporative cooling systems.

Next, look at a very high-efficiency tanked, hybrid heat pump or solar thermal water heater, whichever system is the best match for the type of HVAC system that you have selected. Since you have now significantly reduced the biggest energy loads, the remaining smaller loads now represent a larger percentage of the pie. This is when ENERGY STAR-rated light fixtures, ceiling fans and appliances have the greatest impact.

The next best thing is to improve your ventilation strategies. This is where an energy recovery or heat recovery ventilator (ERV/HRV)—whichever is appropriate for your climate—and humidity control HVAC systems are the best investments. You also might consider installing a separate humidity or dehumidification system if you live in a cold or high-humidity climate, (again, whichever is appropriate for your climate, respectively).

Upgrading features in custom homes allow us to focus more on our individual needs. For some these might result in achieving net zero energy, while for others it might mean a level of indoor quality air that will provide a safe environment for their asthmatic child, and for others it might mean living out their golden years in financial security (at least when it comes to the total cost of housing). In *shades of green* homes, each home may be unique in the combination of strategies used to accomplish our specific goals.

If you or someone in your family suffers from chemical sensitivities or has severe respiratory conditions, spend as much as you can afford to make sure that all of the materials specified for inside the walls or interior finishes are low- or zero-VOC. Additionally, all materials used to fabricate cabinets, trim and flooring products should be specified as made with non-urea-formaldehyde binders. High-MERV filtration provides some removal of dust, pollen and other air contaminants, but central vacuums are also great for exhausting these types of pollutants back outside. However, proper fresh air ventilation and separate HEPA filtration systems are really the only way to address serious air quality concerns. If you plan to install a HEPA filter on your system, be sure that your contractor selects a fan that has the ability to overcome that added resistance and still be able to move enough air to satisfy your HVAC systems needs.

For those whose goal is net zero energy, we recommend smart controls, including smart appliances, smart outlet strips, occupancy sensors on lighting, smart breakers and home energy management systems. These devices allow us to control loads remotely and through automated load detection. For outdoor lighting, solar photocells reduce those loads. Last, but not least, you will select onsite power generation systems that are appropriate for your location. Investigate what your solar power potential is or check local average wind speeds.

If you are in an area where water resources are limited, you should focus on installing a rainwater collection system, if for no other reason than to provide water for your native landscaping. A graywater system is also good for this purpose, and can provide some steady, immediate benefits, as most building codes dictate this water must be used within 24 hours of capture. If your location permits the use of graywater for flushing toilets, it's fairly inexpensive to plumb your bathroom sinks to refill the toilet tank, but it does require some storage device, since typically you don't run the sink faucet until after you've flushed the toilet, and toilets generally won't wait even that long to refill. If potable rainwater is an option for you, this is also a great investment. Installing Water-Sense high-efficiency or dual-flush toilets, as well as low-flow faucets and showerheads can also reduce water consumption. Outdoor living space and native landscaping are also both worthy investments.

However, remember that you cannot achieve this level of green bling on your budget, regardless of how large it is, unless you have followed all of the cost-saving recommendations in the previous two scenarios presented in this chapter. It is those strategies that provide the funding for the green bling that you want. So in addition to those basics, below are the green upgrades that you should consider in building your custom high-performance home.

- High EER/SEER/AFUE/COP-rated HVAC zoned systems, if this is a major energy-using appliance for your climate, with programmable thermostats that measure and control humidity
- High-efficiency water heating system
- Fresh air ventilation system (ERV or HRV)
- Indoor humidity management (if this is appropriate for your climate)
- Improved HVAC filtrations system
- Interior finish materials specified to have minimal impact on indoor air quality and environment, e.g., low-VOC paints and stains, flooring and cabinetry, insulation and subfloor, adhesives, caulks and sealants
- Low-flow plumbing fixtures (approved by the EPA WaterSense program), especially if you are remodeling an older home
- Structured wiring for home automation and energy management
- ENERGY STAR light fixtures and ceiling fans
- ENERGY STAR- or CEE-approved appliances
- Onsite renewable energy systems and emergency power generation
- Native landscapes, rainwater catchment, graywater reuse and high-efficiency irrigation systems, only as required in areas with history of long, extended droughts
- Outdoor living space, including summer kitchen, outdoor shower, outdoor firepit or fireplace

For sustainability of our communities in the future, we need to improve the performance of every home, both existing and those that will be built. Just make sure that whichever level of green you can afford, the home has addressed all of the basic strategies that will provide those base construction cost savings. Above all, you want to…

Avoid Wasting Money

Some products available on the market are either poor quality or not dependable enough to warrant spending your money on. Solar attic vents fall into this category, as do electrostatic air cleaners. Do not allow savvy sales people to talk you into purchasing any green bling without consulting your commissioning agent to discuss their recommendations on how these products will integrate with your overall synergistic house-as-a-system performance. You should also discuss these issues with your homebuilder and have your energy modeling consultant review what return those investments provide.

Common sense must be used to determine what investments will provide you with a worthwhile return. We often hear of money wasted on high-efficiency HVAC systems in homes where little was done to improve the building envelope or right-size them to assure they would deliver the efficiency rating advertised. One of our favorites is the project that adds points for a tankless water heater in a home that requires two or three systems to service each remote area of a poor plumbing design. One of these expensive systems is enough to affect some budgets adversely, and multiple systems are almost assured to never provide even a minimal return on their investment. Although there are some situations where these systems provide benefit, in a high-occupancy primary residence, they may not.

Even those with substantial funds often find that it is not possible to include every green feature and upgrade in their construction budget. So think in terms of what should be done now and what can be done later. This requires thinking about what aspects of the project are easier to do during construction and more difficult or impossible to do later. Finally, think about every purchase that you make and apply the same criteria. Is it durable and easy to care for? Does it require little maintenance? Does it have minimal impact on air quality? Or is it trendy? Make sure you continue to invest in your long-term goals and that you are not sidetracked by products that do not support everything that you have already accomplished.

Adding wood floors, shade awnings over windows and outdoor living space may be projects that you plan to do later, after living in the home for a while, banking your operational and maintenance cost savings to fund them. Just do not be tempted to waste resources (including your money) installing temporary alternatives like carpeting. Just live on the concrete floor—many people find this a fine permanent floor alternative.

Personally, we have seen much success with having a ten-year plan, where you focus on the infrastructure (water, gas, electric) stub-outs for rainwater collection, onsite renewable energy generation or outdoor living spaces during your initial construction. It's much easier and cheaper to do these while the trade contractors are already onsite. Make sure these details are on the house plans so they don't get overlooked. After you are in the home, as your budget recovers from the initial construction strain, you can finish the projects on your own or hire it out. Having all the necessary stub-outs in place will make these "weekend warrior" projects a breeze. Over the next ten years, they should present no strain on your household budget, especially since you will be reaping the rewards of living in an affordable green home.

CHAPTER 10

Keeping It Green

Building a green home gives you a home that is green for a snapshot in time. How the home is used over time is as important to it being green as how it was built. Keeping your home functioning in a high-performance, healthy and durable condition requires your continued attention. To accomplish this, you need a tool kit that consists not of a hammer and paint brush, but of exact product and system upkeep requirements. Every home is a unique blend of products, methods and systems, so maintenance is not one size fits all when it comes to high-performance homes.

If you are a homebuilder reading this book, building or remodeling a green home includes responsibility for educating the homeowners on how to maintain the home in such a way as to continue to reap the benefits of the green features it includes. This education should be presented in person at move-in, with demonstrations of operations and maintenance of all systems and finishes. This applies not only to the initial owners to whom you transfer title, but also to all subsequent owners over the entire course of the life expectancy of the home. Since most homebuilders may not be available to present this information to everyone who lives in the home over that time span, the best way to accomplish this is to create an address-specific homeowner manual. This provides a handy reference manual in a form that the homeowners can look at over time, as they need to address each of the responsibilities required to keep the home performing as it was intended. Since the manual is address specific, the homeowner will understand that it belongs to the house, and not to them personally and will pass it along in future transfers of ownership. Hopefully whoever sells them their next house will have one to give them specific to that address.

The homeowner manual should contain in-depth information regarding every major aspect of the home. This includes specific information about any passive design elements of the home, building specifications and documentation of building materials (warranties and maintenance requirements), a list of all subcontractors used on the job with their contact information, as well as operation manuals for all installed systems and appliances. It is important to include a list of the exact green features of the home, preferably a third-party certified document, such as one from a credible green building program and any related information that's needed to explain the items listed. It is not helpful to give the homeowners a generic list of features, such as those specific to a development, if you have allowed them to choose other options or alternatives. They will need information specific to what is in *their* home.

Educating the homeowner includes communicating the benefits of each product, each system and each strategy used in the construction, so that value is created that inspires them to maintain those benefits. The maintenance checklist should provide details on types and frequency of maintenance needed on all building components and systems. Defined responsibilities include everything from annual walk-around inspections and looking for damage or needed maintenance to providing information to assure that future redecorating and remodeling projects do not undo any key green features of the home.

Here's a suggestion if you are a homeowner reading this book before you begin your building project. Asking potential builders of your project what types of training and documentation they typically provide their clients is a good way to weed out the green-washers from the truly committed green builders. If the response you get from the potential builder reflects only an average quick orientation of the home's features, this may be a reflection on their commitment to green building. Truly committed builders and re-modeling contractors will realize the importance of educating you on high-performance systems and products. They will understand that their long-term success is dependent upon your long-term satisfaction with your home, which leads to repeat and referral sales for them. This chapter is dedicated to the type of information and training that you should expect to receive, in order to recognize the responsibilities associated with keeping your home green over time.

Periodic Maintenance and Repairs

All buildings and building components require maintenance and upkeep. For homeowners to know what that involves, builders should require all subcontractors to provide them with personalized training once the home is complete and all systems are up and running. This is especially true for mechanical systems, including seasonal system operations and periodic maintenance requirements. The builder should also provide training that covers maintenance of exterior building components, weatherization and water management (weather stripping, caulking, gutters, flashings), site features (grading, berms, drains), as well as interior finishes (paint and other finishes, caulking and sealants).

Products do not perform well if not serviced properly and regularly as recommended by the manufacturer. The homeowner manual should include a defined schedule for periodic tasks like painting, back-flushing hot water tanks and changing air conditioning filters on regular calendar intervals. There are entire books on the market that discuss home maintenance, but even these are limited to generic content that may or may not apply to your particular home, or give you enough detail to pertain to your specific materials and systems. Again, this information is different for each home, and your homeowner manual should reflect information specific to yours.

Whether you choose to maintain your home's components yourself or hire it out, you are doing a great service to society by assuring that it gets done. We keep maintenance records for our automobiles in order to protect their resale value. Potential buyers are skeptical if no records exist, and their fears of hidden mechanical problems are often justified. Yet we pay a fraction of what our homes cost for our automobiles. Hiring a home inspector is always recommended, but that does not assure that periodic maintenance has been performed as recommended. These maintenance records should be kept with the original documents and reference materials and made available to future occupants of your home, so the information can be passed from one owner on to the next and the next. It is vital that the required maintenance practices continue in order to pass on the benefits intended by your green features to future occupants and to assure the home is still providing useful service in 50, 80 or 100 years. Some day our great-grandchildren will inhabit these homes, and the condition we have sustained for them will be our legacy.

Commissioning

Over the life of the home, all equipment, systems and materials require periodic commissioning to verify that they are continuing to provide high-performance benefits. This entails performing diagnostic inspections and tests on equipment, systems and home components just like we regularly have our car serviced and its systems checked. Best practices include having a maintenance calendar or checklist that you can refer to periodically to assure that you are meeting each system's upkeep requirements. Some of these tasks require specialized equipment, so you will want to contract with a professional for each type of system. It is beneficial to build a relationship with each technician that you will use and continue with that relationship over time. It is also important that you or your builder familiarize each contractor with how the system integrates with the other high-performance features of the home, so that they understand how the work they will do will affect other systems or the overall operations of the home itself. Sometimes these relationships require monthly, quarterly or annual service contracts.

Lifestyle Choices

Your day-to-day lifestyle choices can play havoc in the performance of the best green homes. If everyone in your family is not committed to the goals that you set forth in energy, water and resource conservation, you may not be able to enjoy the benefits that your home was designed and built to deliver. It is important that everyone understands efficient operations and maintenance and will share in the efforts to keep your home green.

Remember, how we use products and systems can mean more to how green our home is than the products themselves. Using products in the way they were intended can assure years of continued high-performance benefits. If we attempt to manually override design efficiencies, we cannot expect to achieve the intended benefits. This requires that we understand what those design efficiencies are and under what operating conditions they were specified.

One of the best examples of this is the correct use of a programmable thermostat, which regulates the operations of the HVAC system. Most models allow programming for various times of the day and days of the week. That sounds simple enough, but in reality the majority of homeowners do not program it to correspond with their families' scheduled occupancy of the home. Depending on the type of HVAC system installed, this may even require adjusting the program so that the system is able to respond to expected comfort by considering how long it will take to reach the set points desired at each time of occupancy. For example, if you have a heat pump installed, it takes the system longer to achieve a set point of 72 degrees in the winter if it has been reduced to 65 degrees while the adults are at work and the kids are in school. So, if the kids normally get home at 4:00 PM, it is probably necessary to program that higher set point to 3:00 PM in order for it to be achieved by the time the kids arrive. If not, most likely the kids themselves will attempt to manually override the setting and may push the thermostat setting up to 74 or more, thinking that this will get the house warmer faster. In fact this does nothing to increase the speed that the house temperature rises, it only accomplishes a longer run time and less efficient operations.

If this situation happens in the morning before school, it is possible that the manual override will continue throughout the day, long after everyone has gone their separate ways.

Another common lifestyle choice that affects performance is use of exhaust vent fans. Most children (and some adults) do not understand the need to remove moisture and the impact that failure to do so can have on indoor air quality. Teaching these benefits and how to use timers to control the length of operation (to remove the moisture without running so long as to waste energy or create negative pressure) can make this an easy habit change. The same is true for installing motion sensors on light switches or placing a recycling canister next to the trash bin. Making things easier significantly increases their success rate.

You should also pace yourself in making these changes, as too much too fast can result in failure. Create a long-term plan, prioritize which issues are easiest for you and your family to adjust to, which changes fit your scenario and which will make the most impact in your lives. Before you know it, living green will become your norm, and you will feel great about the difference that you and your family are making for a more sustainable world.

Protecting Your Health

Many products that we purchase for use inside our homes release toxic fumes that impact our health, causing everything from allergies to cancer. Be sure that the products that you choose continue to safeguard the health of your family, pets, wildlife and the environment. Other pollutants such as pollen, dust, fertilizers and pesticides enter homes on our shoes and clothing. It's best to have a place to remove shoes at the door, which works as well to protect flooring materials from contamination and wear. Also, make a habit of providing fresh outside air on a regular basis and use the exhaust vents to manage indoor moisture.

Cleaning Products

We recommend that you apply the same considerations when selecting household cleaners and products to care for and maintain your home. Remember, most cleaning products are made with chemicals. Not only are these carried out through our wastewater systems into our waterways and landscapes, but they also leave a residue that is picked up by our skin and noses. The precautions that you take now can reduce the frequency or severity of future occurrences of asthma, allergies and other illnesses.

Pest Control

Use of non-chemical pest control can also reduce indoor pollutants. Diatomaceous earth is one natural organic treatment that can be used on a variety of pests and is not harmful to humans or pets. Many other organic pest control products are available, and many pest control treatment companies have now adopted organic practices.

Home Furnishings

Note that off-gassing is not just limited to interior finishes, it also applies to home furnishings and accessories. Furniture and fabrics brought into your home may also off-gas toxins from wood frames, foam cushions, glues and finishes. Furniture is now available that is certified to use sustainably sourced wood, low-VOC finishes, natural cushion materials and organic, natural fabrics. These materials are not only better for your family, but also are better for the environment and farm workers' health because synthetic chemicals, fertilizers or pesticides are not used to grow them. Look for USDA-certified organic cotton, linen and hemp fibers, labeled as "low-impact fiber-reactive dyes," finished without formaldehyde and machine-washable (to avoid dry cleaning chemicals). Products available include bedding, table linens, kitchen and bath towels, shower curtains and furniture upholstery.

Over time, the bulk of toxic off-gassing runs its cycle, so you do not want to replace products too frequently. Choose products that are classic or timeless in style, as these will be less likely to be replaced than fad styles or "last year's color" palettes. Every time you replace a product in your home, you start the entire health and environment impact process all over again. Typically, salvaged materials or products that have been installed in the home for over ten years may be considered less risky than newly manufactured products.

Hang on to antiques and continue to pass them on—they completed their off-gassing cycle years ago.

Air Cleaners and Fresheners, Incense and Candles

Aerosol or oil air cleaners, fresheners or fragrances release toxic chemicals in the air in order to eliminate or mask the odors. Paraffin candles are petroleum-based products that also release toxic chemicals when burned. The only acceptable scented candles are made from either beeswax or soy.

Another thing to be aware of is the fact that the wick can release lots of soot, lead and particulates into the air.[1] Research has found that some foreign-made candle wicks have a lead wire in them and when these wicks burn the lead goes into the air in your home. It was also found that when wicks are loosely twisted, they create large amounts of sooty black particulates that gets deposited on walls, ceilings, floors and can enter our lungs, too. Wicks should be tightly twisted, have no wire in them and be trimmed to no more than three eighths of an inch in length.

Green Makeovers

Some aspects of green building continue to evolve, with new technologies and products offering higher efficiency and increased durability. Eventually some components of your home will need maintenance, repairs or replacement. Part of keeping your home green is knowing when to salvage and reuse what you have and when it's time to replace something with a new, improved model. So, here are some general guidelines.

Appliances and Light Fixtures

When your appliances are over ten years old, you should consider replacing them with more efficient models. This is especially true for refrigeration equipment, but can also apply to cooking appliances, laundry equipment, light fixtures, heating and cooling systems and water heaters. Be certain to select replacements with the highest performance rating you can afford. Many of these products, including efficient hot water heaters, air conditioners and heaters, qualify for tax credits from local, state and/or federal agencies. If you can, consider drying clothes on an old-fashioned clothesline outdoors in lieu of purchasing a new clothes dryer.

Cabinets

The day might come when you think about remodeling or updating your kitchen or bath. The best cabinets you can put in the new kitchen are the ones already there. New cabinets often mean new wood, which means cutting down more trees, and new finishes usually mean off-gassing. So consider refinishing your existing cabinets. As with all decorating projects, you would be better off if you select neutral finishes and transitional or classic styles that you can update with colorful accents and decorative accessories to achieve a new look so that you'll never have to think about refinishing those cabinets again.

Flooring

Replace carpet and sheet vinyl with hard-surface flooring products that have a long life expectancy. Look for products that are durable, water resistant, low maintenance, easy to clean and made from natural stone or wood products, bamboo, cork, linoleum and marmoleum or ceramic or porcelain tile. You might even skip the materials and just opt to stain your concrete slab.

Landscaping

If you decide to add additional plantings to your landscape, please consult a professional landscape advisor for species that are compatible in size, water and light requirements to those that it will be planted with. Of course, you should always select native and non-invasive species. When adding walkways and patios, use brick, stone tiles, crushed rock or permeable concrete, which allow water to pass through them. Surfaces that allow water penetration mean that rainfall stays in your yard and soaks in, reducing runoff and erosion, as well as reducing your need to water your landscape as frequently.

Successful plantings take about three years to establish good root systems, but once you cross that threshold, they require much less water, fertilizer, pest control and care. Also, aerating annually encourages deep roots and keeps thatch under control. Watering deeply in the mornings will encourage deeper root growth. Plants will be able to go longer between watering, thus becoming able to survive longer through drought periods. Hand watering typically is not done for long enough to accomplish this. Soaker hoses, drip irrigation or subsurface irrigation methods that deliver water directly to the root zone are much more efficient. Make it a point to extend periods between watering to stress the plants enough to encourage their roots to extend out further and deeper, searching for any remaining moisture in the soil.

Remember that you want as much rain and stormwater to soak in onsite as possible, not run off into the street. If you are changing your landscape, make certain it is still graded to drain water away from your home's foundation, and directed into rain gardens for absorption. Since irrigated grass areas typically represent the highest water use in the landscape, consider reducing any current area planted with high-water-need turf species with native drought-tolerant grass species. You can also reduce your lawn area by adding decks, mulched walkways or rock gardens.

Anything you use on your lawn or landscape eventually soaks into our groundwater or is washed off by stormwater into our lakes and streams, which provide much of our drinking water. These contaminants are taken in by aquatic life, including the fish we eat, cause overgrowth of algae and contaminate the water used to irrigate the farmland where our food crops are grown. We do not recommend the use of chemical fertilizers or weed-and-feed products as most of these have been shown to have some cancer-causing concerns. Remember, children like to play on the lawn, and their skins absorb these chemicals.[2] Organic choices can be as effective at controlling pests and weeds while providing rich, nutritious soils for growing beautiful gardens.

Adding organic compost and one to four inches of mulch twice a year can greatly improve moisture retention, maintain more stable soil temperatures, reduce weed seed sprouting and prevent erosion. Preferably, use local shredded Christmas tree mulch, nuisance tree mulch or mulches made of recycled materials. If you find that your mulch is building up and not breaking down rapidly enough, reduce the amount added so that you are just replacing what has gone back into the soil. If mulch becomes too deep, it blocks the penetration of rainfall. Too "mulch" of a good thing is bad.

Grass kept at least two and a half inches tall is healthier, has deeper roots and will provide a thicker turf. Design your landscape to keep your turf area to a minimum and use a push lawnmower or an electric model that is much less polluting than gas-powered types. Whatever type of mower you use, be sure that it mulches the grass so that it readily breaks down to feed your lawn, inhibit weed growth and reduce the need for additional fertilizers and herbicides and more frequent watering. Recycle that polluting, energy-using leaf blower and buy a broom.

With a native plant-based landscape, it is important to provide fresh water resources for the wildlife that will inhabit it. Also, add hummingbird, bird and squirrel feeders and some bird houses. Plant annual wildflower blends to attract butterflies. Add a bird bath and goldfish or koi ponds for special interest. We each can do our part to reduce the strain on the wildlife suffering from loss of natural habitat due to our sprawling community developments. Remember who was here first and let them know they are still welcome. We cannot survive on this planet without other species, so we need to provide them space to coexist with us.

Plant deciduous vines on these arbors or trellises to provide shade in the summer months. These vines will then lose their leaves in the winter to allow the sun's warmth to pass through. The sun is lower toward the southern horizon in the winter, so these types of architectural features allow the sun's heat to enter through the windows in colder months. Controlling solar heat gain with these methods is preferable to interior window treatments like blinds and draperies because it stops the heat before it enters the window.

Light Fixtures

Older light fixtures and bulbs can be big energy users. If you are re-modeling an older home, consider replacing old recessed can lights with the new LED fixtures. The bulbs can last 20 years or more and use a fraction of the energy of older incandescent bulbs. These new bulbs are available with great color spectrums. ENERGY STAR-rated light fixtures and ceiling fans are much more energy efficient and only accept these newer types of light bulbs. Even if you do not replace all of your fixtures, you should at least replace incandescent bulbs with compact fluorescents (CFLs).

Plumbing Fixtures

Again, if your plumbing fixtures are from a pre-water-conserving generation, consider replacing them with low-flow models. The green dollar payback is quite reasonable for most of these replacement models. If you decide not to replace your fixtures, by adding low-flow aerators to your kitchen, utility and bath faucets, you can reduce water usage without affecting perceivable water flow or pressure. Look for aerators with a flow rate of 2.2 gallons per minute or less.

Older toilets could use 3.5 to 7 gallons of water per flush (gpf), while new federal limits on standard toilets is 1.6 gpf. WaterSense-rated toilets use even less. A highly regarded website that provides guidance in toilet performance is the Maximum Performance (MaP) testing program (map-testing.com). MaP is a voluntary testing program sponsored by a large number of American and Canadian water and wastewater utilities and governmental commissions. MaP tests toilets repeatedly to find out how many grams of waste each one can reliably flush and at what water use. These results are all published for you to use in selecting your best fixture.

The US EPA WaterSense program requires that all WaterSense toilets meet a minimum MaP threshold of 350 grams. If you wish to be sure that a "plumber's friend" (plunger) is not a part of your future, you should know that many toilets on the MaP database successfully achieve 1,000 grams per flush at WaterSense flow rates.

Shading Devices

Consider adding shading to reduce the solar heat gain on your home. Awnings, trellises, arbors or other exterior components, especially on the east and west sides of your home, can help to block the sun's heat in warm months. Plant deciduous vines on those arbors or trellises to provide seasonal shade. These vines will then lose their leaves in the winter to allow the sun's warmth to pass through. Controlling solar heat gain with these methods is preferable to interior window treatments like blinds and draperies because it stops the heat before it enters the window.

In fact, using external shading devices to provide seasonal control of solar heat gain into the living space is preferred to replacing old windows with newer ones carrying a low NFRC SHGC (solar heat gain coefficient) rating.

You should only replace windows that are leaky, i.e., causing water damage to your building assembly, or that the window assembly itself is leaky, causing significant infiltration (heat loss or gain).

Wallpaper

New non-woven wallpapers are vinyl-free and breathable. They are also tear-resistant for easier installation and removal later, and even reusable. Better yet, shop for non-woven papers made from wood pulp that come from an FSC-certified forest and use water-based dyes and pigments.

Window Treatments

If your lot size limits your ability to add any type of exterior shading, it would be good to install drapes or blinds on unshaded windows, as this can also reduce your cooling and heating bills. If you have a west-facing room with several windows, closing the drapes or blinds can make the difference between a room that is tolerable and one that isn't. In heating climates, the right window treatments can act to insulate you from severe outside temperatures and reduce heat loss through the highly conductive glass. And, of course, if you need interior window treatments for privacy, you should try to raise or open them during the daylight hours in winter to allow more solar warmth and natural light into your home.

Green Power Alternatives

The term "green power" is the general term used to refer to energy from all renewable energy sources, including solar, wind, hydroelectric, geothermal, biogas and biomass waste. Many electrical utility providers offer green power as an alternative to purchasing power generated from facilities that burn fossil fuels or from nuclear power plants. This green power is not dangerous to produce and does not contribute to pollution, global warming or toxic waste that contaminates our environment.

In your area, you may notice large arrays of solar photovoltaic panels, large wind turbines, hydroelectric plants or methane gas generators near landfills. About 50 percent of all retail customers in the United States now have an option to buy green power from their local electric service provider.[3] To find out if you have this green power choice, contact your local electric service provider or visit www.epa.gov/greenpower.

Another option for purchasing renewable energy is to purchase renewable energy certificates (RECs).[4] An REC represents one megawatt hour of electricity generated from renewable sources. These certificates are available for sale separately from the associated electricity, so they are not the same as green power. You can purchase these certificates from numerous renewable generators, both within and outside of your own electric service area. By purchasing RECs, you are supporting the production of green power even if it is not available in your area.

Getting to Zero
Looking into Our Crystal Ball
Toward the Future So That Our Current
Decisions Might Stand the Test of Time

CHAPTER 11

The Zero Energy Capable Home Model

There are many ways to reduce grid-supplied energy use in a home. These efforts run the range from extremely expensive high-performance systems to others that cost nothing at all, resulting completely from increased conservation. It does not make any sense to spend more on the systems used to reduce energy use than the cost of the energy saved, although we have seen very large, expensive solar photovoltaic arrays on mansions built with little thought given towards efforts to reduce consumption. Nor is it reasonable for homebuilders to install energy control devices without a full understanding of the home occupants' lifestyles or with the expectation that those using these devices will make major changes in their own behavior in order for them to work as intended. The result of these misguided efforts is almost always failure. For example, we have seen homes built with water heater recirculation systems on timers that the occupants had reset to run 24/7. This was done because at some point someone in the house needed hot water outside of the programmed operation of the timed settings. They did not understand the energy-saving benefit of the timer and therefore had no reason not to modify its operation to allow continuous circulation so that situation would never repeat itself.

A realistic approach is somewhere in the middle of this. Home-builders should be only be installing high-performance systems and appliances that have reasonable payback on investment, while at the same time making an effort to improve how users understand these systems. It is important for occupants to learn *all* the benefits of efficient operations and how their own behavior affects those results.

You might wonder if this middle-of-the-road approach is enough to get close to net zero energy. Of course the fastest way to net zero is to live like a scrooge, freezing in winter and sweating all summer, unplugging appliances not in use and to go about all activities in the dark except when task lighting is required. If this were the mandate promoted for energy conservation, few people would participate. We certainly wouldn't. What is needed is a reasonable, cumulative approach that provides an optimum combination of the strategies discussed in this book, including those related to how the home will be used. It starts with design and ends with smarter homeowners, and very little is missed in between.

Net Zero Defined

For those of you who need clarification, a "net zero energy home" is defined as a home that operates efficiently enough that over the course of a calendar year it produces as much energy as it uses, resulting in no additional energy purchased from outside sources (i.e., your public utility provider). This is *not* an off-grid home.

During milder seasons, the onsite energy production should exceed the amount of energy the home consumes, so it is able to bank its excess on the public utility grid (instead of storing it in batteries), accruing credits with the utility service provider. But if, in more extreme temperatures, the onsite production is not able to produce enough power to handle your heating and cooling loads, you draw on your stored reserves and purchase supplemental power from the grid as needed. Since this home depends on the public electric grid to store the excess energy it produces, it achieves net zero at the end of a consecutive twelve-month calendar period, when the amounts sent and drawn from the grid zero out. In other words, you have not purchased any more kilowatt hours than you banked during the year. Usually a net zero energy home's onsite power is generated by some combination of wind power, solar photovoltaic (PV) and solar thermal system. Other sources are available, but less common. These include hydroelectric power, steam, bio fuels, wood or fossil fuel-burning generators. Of course, the latter of these (wood and fossil fuels) would not be considered green energy sources.

Bear in mind that many electric utilities use net metering, which means they bill only the net kilowatt hours (however much more is consumed than generated onsite). The billing rate is the normal energy charge per kilowatt hours for the excess consumed. However, if you generate more than you consume, they may only give you monetary credit at wholesale electric rates. So, again, net refers to the difference between kilowatt hours produced and consumed over this period.

To further clarify, a zero energy *capable* home is the same concept, but does not have an onsite power system installed. It just means that the home was designed and built efficiently enough that if you were to install a reasonably sized onsite power generation system, it could achieve a net zero relationship to the grid. For many homes on heavily treed lots or in densely developed communities with tightly packed neighboring buildings (creating wind breaks or shading roofs), net zero energy capable is the best we can hope for. These scenarios are not supportive of installing solar panels and wind turbine installations. If you are interested in having a solar survey done of your home or building site, many companies that sell solar PV systems provide that service, as do electric utility companies that support incentives for these installations.

Our goal, for the purpose of this chapter of the book, is to achieve a net zero energy home. This means net zero energy over a consecutive twelve-month calendar period using the grid to manage our electrical production and consumption, or, in essence, using the grid as our storage device. To do this, we must reduce our annual usage to a level that can be produced by some type of onsite power generation system that is reasonably sized for this home, can be expected to provide our average annual electric consumption and gives us a worthwhile return on that investment.

To represent the cumulative effects of the strategies that were presented in the first ten chapters of this book, we will start with a typical home built to base building codes, with nothing special about its design, materials, construction methods or systems. We will then apply only such basic changes to its design that allow us to develop a baseline that we can use to measure the improvements in performance from our recommended better-than-code materials, construction methods and systems, including an onsite power generating system to supply its annual energy needs. The same strategies would be applied to achieve a zero energy capable home, only no onsite power system would be installed.

The Typical American Home

First, let's define the typical home that we will compare this to. You might be surprised to learn that tiny homes do not enjoy reductions in energy use equivalent to their size. Even though smaller homes are more energy efficient than larger ones, their power density per square foot is higher. The problem is that all homes have a kitchen, and even the simplest kitchen generates base appliance loads. It is not really reasonable to exclude laundry equipment from a household; even if your design places them outside of the thermal envelope, they still generate electrical loads. And your home is probably going have at least one indoor bath, one water heater and one bedroom area. So, even the smallest homes have these base electrical, occupant (ventilation) and heat loads. Small homes also have the same building code-required door and window openings for safe exits in case of emergency that are required in larger-roomed homes. This means that window-to-floor area ratios can actually be greater in small homes. Given all of these considerations, it is much more difficult to improve energy efficiency in a small home than in a larger one.

For the purpose of creating this net zero energy model, and for the sake of argument, we are using the average family home size built in the US that has 2,400 square feet of conditioned living space.[1] This home would not have given any thought to efficiency in design, materials or systems, other than those required by the 2009 International Residential Code (IRC) and 2009 International Energy Conservation Code (IECC). Note that, at the time of writing, this is the worst house that you can legally build in many areas of the country, meaning that this home only meets legal code standards. Unfortunately this is the quality of construction that is typical of many homebuilders.

Some people mistakenly think that building codes represent an exceptional level of quality. In fact, they are the minimum baseline for all construction. Think of it this way: if you build anything below code, they stop your project until you improve it to the minimum acceptable level.

For the purposes of our comparison, this typical American home is a two-story home, including three bedrooms and two and a half baths (with a half bath downstairs). It would have a standard 2-car attached garage (approximately 400 square feet), giving it approximately 2,800 square feet under roof. It has a chopped-up floor plan and roof design with lots of windows. It was not designed with any thought given to resource efficiency. Its building site was selected based only on its "street appeal." Additionally, no thought was given to the home's orientation for passive benefits, nor for the ability to add solar panels to the roof. In Austin, Texas (home to the writers of this book), such a home built with R-30 insulation in the attic floor, 20 percent glazing (glass) to the conditioned floor area, an electric 40-gallon tank water heater, and electric furnace and air conditioner consumes around *19,352 kilowatt hours* annually.[2]

Our Baseline Home

It is necessary that we improve both the design and features of the home in order to achieve our net zero goal. Trying to compare our net zero home to the typical American home is like comparing apples and oranges, so we need to come up with a basic home design that we can model against itself in order to see how each of our strategies improves its performance. So our first improvements to our typical American home are to create the baseline energy model so that we compare each material, method and system performance to the same basic floor plan (identical footprints) and even some of the same types of features. In fact, the IECC requires that the HVAC systems used for the baseline home and the energy-efficient improved version (in our case, the zero energy home) are the same basic type. In our case, we are using heat pumps. Our location will be the same for both the baseline and our zero energy homes—outside of Austin, Texas.

To develop our baseline home, we are going to look at improving the home design. We start by looking at the right size of home to fit our needs. Keeping the bigger picture in mind, we think about our long-term needs and come up with a home that is designed to morph over time as we expect our needs to change. We want a home that we can initially raise a family in, but that will not be too large after the kids are grown and gone. More on this type of design will be discussed in Chapter 15.

Because we want to follow the "Not So Big House"[3] design considerations discussed in Chapter 2, we are going to start by reducing the conditioned space of the home. Since our goal is net zero energy, we want to design a structure that will allow us to maximize the thermal envelope. So we design a basic rectangle, on two-foot modules. This simplifies the design to a 40-foot-wide × 28-foot-deep footprint, which gives us an improved shape factor (foundation perimeter^2/foundation area) of 16.5 and reduces our conditioned space to 1,816 square feet.

The home is a three-bedroom, two-bath and has a cooking appliance, a dishwasher and refrigerator, as well as an indoor washer and dryer, just like our typical American home. However, unlike that home, it has a detached open-air carport in lieu of the attached two-car enclosed garage. This not only reduces our construction costs, but also improves our indoor air quality by separating the garage source pollutants from our living space.

For modeling purposes, we are using the 2009 IECC codes for both the baseline home construction specifications. The wall framing is typical wood construction, with overuse of studs at intersections and openings by inappropriately (but typically) trained workers. The walls and floors will have fiberglass batt insulation installed to an ENERGY STAR Grade III quality (the unfortunately typical level of gaps, voids and compressions). The ceilings will have R-30 blown fiberglass. The windows have a U-value of 0.65 and solar heat gain coefficient (SHGC) of 0.30. It is an all-electric home.

This baseline home has a 13-SEER and 7.7-HSPF heat pump split system air conditioner/furnace in the ventilated unconditioned attic space. We made the decision to install a heat pump rather than a gas furnace because gas furnaces on the market today are not very efficient, even those with higher AFUE ratings (i.e., 95 percent efficient). In energy modeling, if we convert gas and electricity to the same units of measurement, high-efficiency electric heat pump systems always easily win hands down.

According to ASHRAE ventilation defaults, a three-bedroom home is considered to have four occupants, one in each secondary bedroom and two in the master bedroom. For water usage, we used the IECC 2009 "Specifications for the Standard Reference and Proposed Designs" baseline of 30 gallons a day plus 10 gallons per bedroom, or a total of 60 gallons per day. For water heating, we have installed one 40-gallon electric water heater, rated at 0.92EF. The lighting is incandescent, and it has standard builder-grade electric appliances.

So, essentially, in our baseline home we have:

- Reduced the conditioned space of the home and detached the parking structure
- Simplified and improved its footprint and its basic design
- Upgraded to an air-source heat pump HVAC system, in lieu of the typical American home's electric-resistance or oil-burning furnace

- Met 2009 IRC and 2009 IECC code minimums for all systems, construction specifications, lighting and appliances

To finalize our baseline home design we need to examine the effect of its windows on our annual consumption. The typical American home, from the perspective of a building designed for a typical production homebuilder, has window placement for architectural balance on an otherwise drab box design. This home has minimal code-required overhangs and no shading devices to accomplish the desired goal of letting in lots of natural light, again to add interest to otherwise drab interior finishes. When we tallied up the window sizes and placement for this scenario, we found the total window glazing to floor area is about 18 percent. For most energy efficiency programs, the threshold recognized for limited glazing in a design is 15 percent, yet in the real world, we see many homes built with nearly full-glass walls, far exceeding this recommendation. Yet, even without reducing our glazing area below 18 percent, we were able to reduce the home's annual energy consumption to *16,267 kilowatt hours*. What a difference the reduction in home size and shape factor improvements made! Even with our smaller home's power density penalty, we still reduced our energy use by 16 percent compared to the typical American home.

Again, since our goal is net zero energy, we must reduce the glazing area in order to improve the overall thermal performance of our wall assemblies. We want (and need) window placement for daylighting, passive ventilation and views. So now we are approaching glazing from a functional perspective, so that we can actually use windows for the benefits they can provide, not just as architectural elements. As a result, we were able to reduce our total glazing area to around 159 square feet, which is about eight percent of our conditioned floor area. This revised approach to glazing reduced our total glazing area by more than half!

We can add this feature to our baseline specification:

- Reduced total glazing area to 8 percent of conditioned floor area without any shading devices or improvements to windows ratings

The result of reducing glazing alone, even with a worst-case orientation, reduced our annual energy consumption to *15,646 kilowatt hours!* Those 621 kilowatt hours of savings may not seem like much, but considering that we have not yet made any improvements to reduce the heat gains in our building envelope, it's a good start. This is our baseline annual energy consumption defined by design and building specifications that we can use for comparison to measure the construction and system improvements that will achieve our net zero energy goal.

Net Zero Energy Home Model

Now our goal from here on is to look at all the things that use energy in the home and determine how we can make each of them as efficient as possible. We want to get our energy consumption down to equal to or less than what we can produce with a reasonably sized PV array so that we achieve our goal of net zero energy at the end of a consecutive twelve-month period. In order to accomplish this, details are important. We will leave no stone unturned. We'll start by examining more aspects of the home's design as well as construction specifications, then we'll move on to improved building materials and methods, then improve the major systems that use the most energy, then to those electrical gadgets that provide us with the amenities we expect in modern homes.

But, to further clarify, *our goal is to achieve net zero energy at a reasonable overall cost.* We define this as a cost that realizes a payback of the incremental investment in an acceptable timeline. We must look at the house as a system, with each design and construction decision based on how much impact we have in total performance, in other words how well the individual components and strategies support each other and provide overall benefit. To do this, we must assemble a team of committed experts, from all our major trades. Typically this team would consist of an architect with experience in passive solar design in our climate and a homebuilder with established relationships of trust with his major trade contractors. These trades would have expertise in high-efficiency products and systems, including structural framing/envelope details, insulation, windows, HVAC, water heating and lighting design integration. The benefits of this team dedication to achieving the project goals cannot be stressed enough.

The MacLeamy curve, introduced in Chapter 3, shows that we get the greatest return on investment from design improvements, and next with improving the building envelope. So, to start with, our net zero design will actually increase our area under roof to 2,651 square feet (including the 400-square-foot carport), but with the same reduced conditioned space of 1,862 square feet. By adding 346 square feet of covered porches and balcony for outdoor living space, we end up with 2162 square feet of total living area. We determined these changes by looking at the building site and determined the best case orientation (the modeling software tells us what this is, in our case south) in order to achieve the best overall energy reduction. This model showed us the benefits of adding a covered front porch and first- and second-floor side porches. These shading devices helped us to mitigate sun exposure on the east and/or west sides, and use passive solar strategies on the south-facing windows to shade them in the summer while allowing heat gain in the winter. Note that, if you have a lot that faces west, you can still design the home with south roof solar access and minimized solar exposures on the east and west sides by placing porches, carports or shade awnings on those sides of your design.

This outdoor living space gives us plenty of room to add a summer kitchen and dining/sitting area for entertaining. Having these additional areas under roof as non-conditioned space not only reduces the amount of resources and money needed to build them (wall framing, insulation and interior finishes), but also saves resources over time for maintenance and operations. If you live in a predominantly hot climate, but occasionally enjoy a few cold winter days and you decided that you needed a fireplace, you would add it in that outdoor living space, improving the overall thermal performance of your home, reducing conditioned air losses up the chimney, as well as eliminating that source of indoor air pollution. If you live in a predominantly cold climate, you would design and place these outdoor living spaces to shade your south-facing glass in summer but allow the warm sun to enter your home from fall through spring. This is also the best side of your home to have a four-season room if you so desire.

Austin, Texas, where we happen to live, has great solar access, so to achieve our net zero energy, we know that we want to install a solar photovoltaic system as our source of onsite power generation. It is at this point that we must redesign our roof to accommodate the solar array. Our solar designer can guide us to ensure that the pitch of our panels for our latitude maximizes electrical production from the system. We will also need to optimize our roof space by making certain that it is unobstructed (not shaded by trees, adjacent buildings or other roof elevations) with good southern solar exposure, minimal (or no) penetrations for plumbing vents, fireplace chimneys or anything else. By working with your plumbers and other trades early on in the process and making them aware of these needs, they can plan things like vent placement to avoid the solar panels.

For our 1,816 square foot of conditioned space, a reasonably sized system would be about 6 kilowatts, which is capable of producing 8,169 kilowatt hours a year based on our design and installation. This is a common size for a solar panel array, and it will fit nicely on the south-facing roof area that we have designed to accommodate its installation. Of course we might be able to achieve net zero energy with a *much* larger solar array without making any energy improvements to the building, but that is just throwing good money after bad, which is not in keeping with the spirit of this book.

We also made some interior floor plan modifications, placing the kitchen on the north side of the home and all of the bedrooms upstairs on the south side, with their windows designed to capture the prevailing southerly breezes. We looked more closely at the design for necessary revisions to the plumbing layout for efficient hot water delivery and HVAC layout for efficient delivery of comfort. We then placed all wet locations within close proximity to each other so that they can be easily serviced by a centrally located hot water storage tank. We have centrally located the air handler and furnace and shortened the duct runs to terminate at the interior walls of rooms for the most efficient delivery of air.

Of course, we are going to make improvements to the construction materials, methods of construction and efficiency of systems in our green home, according to the recommendations of this book. These improvements will also support our net zero energy goals. To start, we need to look at our building specifications to make certain we are reducing the amount of energy needed for mechanical heating and cooling for comfort within the conditioned space. We do this through attention to the quality and performance characteristics of our framing, insulation, windows and sealing up of the building envelope. Failure to pay attention to these details at this point can result in energy penalties that will impact operational costs throughout the life of the home.

Net Zero Energy Home Building Specifications

- 2 × 6 wood-framed wall using Advanced Framing (OVE) techniques
- Open-cell spray foam for improved thermal envelope performance of the R-15 "total" cavity fill wall insulation
- Half-inch R-3 rigid board insulated sheathing
- Sealed attic with spray foam under roof deck (R-21)
- Tight construction with continuous thermal and air barriers installed
- Thermally broken windows with a U-value of 0.30 and SHGC of 0.25 (for our southern location—this would be U-factor of 0.25 and an SHGC of 0.45 to allow solar heat gain in winter in the north)
- Windows designed for daylighting and passive ventilation
- Vented metal roof with an SRI of 0.35
- Slab-on-grade foundation with one-inch R-5 rigid board insulated edge (or, in other climate zones, the R-value required by IECC code in your area)

- Insulated exterior doors
- No skylights installed

We used resource-efficient design details that will also permit us to improve the energy efficiency of the envelope (Advanced Framing = improved insulation). We want to make sure that we have continuous and contiguous thermal and air barriers installed and that we seal up every penetration in the building envelope. Moving the insulation from the attic floor to the roof with spray foam insulation at the rafters (or Structural Insulated Panels) brings all of the mechanical equipment (HVAC air handler and furnace) and ductwork within the thermal envelope. The metal roof gives us a reflective surface to shed the hot summer sun (we are located at a latitude that has direct solar exposure on the roof in the hottest part of summer). It also provides us with a good surface for mounting solar panels and collecting rainwater, and we should expect to see a reduction in our home insurance costs due to lower fire hazard.

We optimized the benefits from our southern-facing windows for passive solar heat gains in the winter and provided shading devices (covered porches or awnings) to protect them from heat gain in the summer. Having reduced the number and size of windows on the east and west walls (to a total of 30 square feet on the east and 14 square feet on the west) and having added shading devices, we have really limited our solar heat gains. We engaged the services of a building envelope commissioning agent to analyze how this building might perform over its life expectancy in terms of continued efficient operations. We want to achieve net zero energy over the long term, so we need to be aware of what conditions could impact performance over time. Using these strategies in our energy model, we realize *a reduction in annual energy usage to 13,897 kilowatt hours.*

Even though this energy reduction may not seem significant, we probably would never achieve net zero without making these improvements to the design and the building envelope. Also, making these improvements may indeed be the most important initial efforts that we make, as these are the most difficult to remedy or retrofit after construction is complete. Next, we are going to look at the mechanical systems that use the most energy in the home's operations.

HVAC System

It is important to note that, in modeling the home *during the design phase*, we made revisions to the design and building envelope components in order to attempt to reduce the heating and cooling loads. To assure its operational efficiency, we want to reiterate that we started by designing an efficient system: central equipment location and short, straight duct runs.

Now our mechanical contractor uses ACCA Manual J software, version 8 or later, to correctly calculate how many BTUs of heating and cooling each room in the home requires. Following ACCA recommendations, we are designing for HVAC capacity at 96°F in the summer and 30°F in the winter (you, of course, would adjust these as recommended for your climate). These temperatures are at the most conservative end of the ACCA-recommended design temperatures for our area. A little-known secret about ACCA Manual J is that it contains "fudge factors." The load estimates derived from Manual J are about 20 percent greater than the true load on the house. This is done intentionally so that, when extreme outdoor temperatures occur, the system will still be able to achieve the thermostat set point temperature without it being extremely oversized the remainder of the time.

Note that the software provides minimum input for loads related to lighting and appliances if the contractor wishes to make those entries. It is more difficult to model or foresee miscellaneous electrical loads from cable boxes, printers, sound systems and other home electronics. So we are basing our sizing on the home's design, construction inputs and occupant loads. With our climate, design and construction specifications loaded into the Manual J software, the resulting report recommends a two-ton system to meet the load needs of our home. By entering the load components separately specific to the first and second floors, the software is able to further discern that we need 0.7 tons for the downstairs area, and 1.2 tons upstairs.

In our typical American home, it would be business as usual for the mechanical contractor to size the system strictly on a rule of thumb calculation of 350–500 square feet of conditioned space per ton of cooling capacity and then round up. So, in our 1,816 square foot home, this would have meant installing a 4- or 5-ton system. Not only are those systems more expensive to purchase initially, they would also operate very inefficiently. Short-cycling would have surely been the result. This would significantly increase our monthly electric utility bills, reduce our comfort due to high indoor humidity, and also shorten the life expectancy of the equipment.[4]

Based on the load and airflow needs indicated in the report, Manual S calculations are performed in order to select equipment that is capable of achieving the desired thermostat set point for both the seasonal cooling or heating loads with a blower capacity that can move the necessary volume of airflow to effectively distribute to all areas of the house. If we do not balance the system's capacities to match the needs of the home, we end up with a system sized to fit one need and oversized or undersized for the other. This is sure to result in excessive energy use and poor comfort.

Next, Manual D procedures are used to correctly size the ductwork and design it to efficiently deliver the right amount of air and BTUs to each room using the selected blower. This assures that we will be able to maintain comfort in all seasons. Most comfort complaints are not due to undersized equipment, but to incorrect and inadequate airflow.[5]

In most climates, the HVAC system is the biggest energy hog in the house. For this reason, we decided to install an 18-SEER (air conditioning efficiency) and 8.9-HSPF (heating efficiency) split system heat pump. If you are in a winter-dominated climate, you should maximize your heating efficiency first and not be so concerned with the air conditioner efficiency. All interior system components (air handler and ductwork) will be within our thermal envelope. We also planned to make sure that we seal up the duct-work and that we test for duct leakage. In addition it is crucial for your comfort that the airflow from every grille in each room be measured and matched to the Manual J. It is also critical that system total external static pressure be measured and that it does not exceed manufacturer specifications.

Having a system that works on paper is great, but having one that works as installed in your home is better still. We should make certain the mechanical contractor has performed a proper start-up procedure and adjusted the refrigerant charge for the needed airflow. Only by doing so can we verify that the system is delivering its stated capacity. To this end, we recommend that your contractor perform the ACCA Quality Installation procedure at system startup.[6] This is the only way to know you are getting what you have paid for.

We are also including an energy recovery ventilator (ERV) in our home, since we live in a hot climate, to bring in fresh air and exhaust stale air. Systems like this are generally referred to as mechanically controlled ventilation. In our climate, these systems are about 50 to 70 percent efficient at reducing the heat and humidity load inherent with ventilation air,[7] but that is better than nothing, so we'll take what we can get. If you live in a drier or more heating-dominated climate, you should consider a heat recovery ventilator (HRV). These units warm incoming winter ventilation air but do not control summer humidity between the incoming and outgoing air streams. Since they do use fan energy to improve indoor air quality, they negate some of the savings we have achieved from our high-performance equipment, reduced internal loads and more efficient operations. Even then, the modeling software indicates that these system improvements have reduced our total home annual energy usage to *12,988 kilowatt hours.*

Water Heating

The next big mechanical system that we need to look at is water heating. Water heating often consumes about 20 percent of the total energy used in a home.[8] We're replacing the electric water heater that would be standard in our all-electric home because of the inherent high cost of heating water with electric resistance, but also because it has large standby losses in storage. The system we are going to install is solar thermal, using the abundant energy we have available from the sun to do the primary water heating. Of course, this also should be something we plan for in design and construction, as these systems generally require a larger storage tank (that houses the supplemental backup electric water heater), as well as supplemental piping, reinforced roof framing and controls, which makes it more difficult to retrofit later. For reasons that will be discussed in Chapter 15, we are installing two separate storage tanks, one on each floor, serviced by solar water heating panels on the south-facing roof.

We are also making sure that all hot water lines in the house are insulated, and again we have centrally located the systems with short line runs to each wet location. An acceptable alternative would be to install an on-demand water recirculation system on each structured plumbing loop. We plan to have low-flow faucets and showerheads, as well as a WaterSense-rated dishwasher and washing machine. These improvements result in a reduction in our annual energy usage to *10,656 kilowatt hours*. This alone has been the single biggest improvement we've seen so far.

Appliances and Lighting

Incandescent lighting not only uses considerably more energy than its high-efficacy—CFL and LED—alternatives, but also generates enough heat to add unnecessary heat loads to our home's cooling systems. As our main systems (HVAC and water heating) become more efficient, the loads generated by lighting and appliances represent a larger percentage of the total reduced load. We can really make an impact here.

We are specifying ENERGY STAR high-efficacy light fixtures and bulbs in 90 percent of our interior and exterior fixtures, as well as installing ENERGY STAR-rated ceiling fans and bath exhaust fans throughout the home. Our lighting consultant will provide a design that gives us a mixture of fixture types and locations to provide the type of lighting we need, where we need it and with control capabilities to match use. Some fixtures (bath exhausts) will be on timers, while others (lighting) will be on occupancy sensors.

ENERGY STAR appliances have gained market share to the point of being mainstream. But as we mentioned in Chapter 6, there is a new level of appliance on the market, Consortium for Energy Efficiency (CEE) Tier 3. Refrigerators and electric clothes dryers top the list of energy users in the home. CEE Tier 3 refrigerators are 30 percent more efficient than federal standards,[9] and any cost premium is easily offset by the energy savings. Additionally, we have selected an electric induction cooktop, the most efficient cooking appliance currently on the market (short of a solar oven), efficiently converting 90 percent of the energy used to heat.[10] For the cooks out there, it also gives instantaneous and reliable temperature control. We also plan to install a clothesline in our net zero energy home, so we expect to reduce the use of our electric dryer by 50 percent. This move gets our annual energy usage down to *7,764 kilowatt hours*, well worth the investment. With these improvements, we have already exceeded our goal; we are well within the range of what a six-kilowatt solar system can produce. But why stop now? Let's look at what else we can do.

Influencing Operational Efficiencies

Miscellaneous electrical loads (MELs) can be almost as high as the lighting and appliance loads (which include entertainment and computer appliances) in our home. In fact, our model indicates 2,314 kilowatt hours a year! We have seen the development of more smart appliances and controls on the market in the last few years. Computers can now sense that they are not active and automatically go into sleep mode, and the same technology is now being applied to televisions and video equipment. Smart outlet strips that sense no demand can cut off phantom loads to appliances plugged into them, yet have a couple of non-switched circuits to keep things like cable modems or clocks from being controlled with the rest of the batch. The first generation of whole-house smart energy management systems includes smart grid technologies, smart meters and smart electrical panels and breakers and more.

The only thing left is occupant behavior modification. We said that like it was no big deal, but perhaps it is the most significant savings yet to be analyzed. In the future, we will most likely see significantly higher electric energy costs, especially during peak demand periods. Just like you might consider not running errands in your car during rush hour traffic, minimizing your energy use during peak demand periods may be something you will have to think about in the future if you aren't already. Just like you program your thermostat to a different (i.e., more energy-conserving) setting while you are away at work, you might program your dishwasher to run during the night, when electric rates are lower.

Since we are installing a solar photovoltaic system on the house, we found a contractor who offered us energy monitoring as part of the purchase package. With this, you can monitor your energy usage in real time, which in and of itself makes you more keenly aware of leaving on lights in rooms that are not being used, high thermostat settings at times when everyone in the household is off at work or school and even those phantom loads consumed by electronics that are not being used. This awareness helps you modify your behavior and gives you something to show off to family and friends, increasing awareness all around. It also encourages investing in smart controls, like occupancy sensors and smart outlet strips, to help accomplish lowering your usage. In some neighborhoods, where these energy monitors have been installed on every home, we are actually seeing neighbors compete for the lowest utility bills. This is competition as a good thing!

Clearly, with little effort, awareness and behavior can be improved to contribute to net zero energy reduction. We believe that with a little attention to details, MELs can be reduced by 15 percent

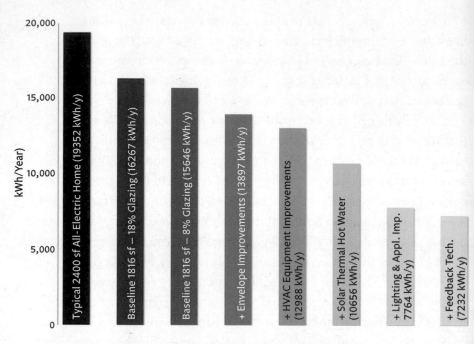

FIGURE 11.1. Energy model graph.

in our net zero green home.[11] Taking one last look at our model reflects that we are at *7,232 kilowatt hours a year*, having exceeded our net zero energy home goal.

We have achieved our goal! "Houston, the Eagle has landed." Congratulations. And we did it without any crazy, extreme or prohibitively expensive methods. Affordable-$$-Healthy-High-Performance Green Lives!

In case you are wondering how we did these calculations, there are a few computer software programs on the market that have been developed for this purpose. We used EnergyGauge, developed by the Florida Solar Energy Center, but most HERS (Home Energy Rating System) consultants use REM/Rate, a similar software developed and licensed by RESNET, the Residential Energy Services Network. These programs are both approved by RESNET for use by the mortgage industry in qualifying funding under energy-efficient mortgage (EEM) programs. There are other government-sponsored whole-building energy simulation programs on the market, includ-

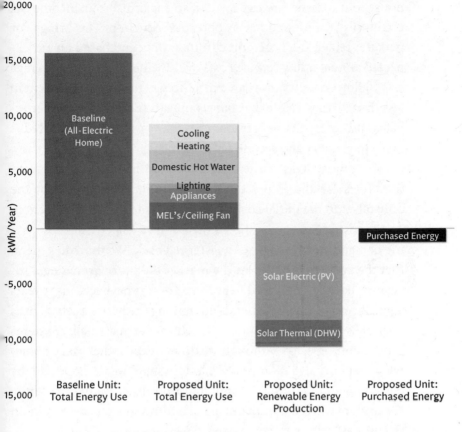

FIGURE 11.2. Excess solar production.

ing DOE2 and Energy Plus, developed by the US Department of Energy.

The Solar Photovoltaic System

Note that, based on the current market price, we expect to pay a little over $3.00/watt for the solar photovoltaic system installed. That's about $19,000 gross cost. However, we are fortunate to live in an area serviced by an electric utility provider that offers a rebate for some of that cost, up to 6,000 watts. We'll max out this kilowatt limit, resulting in a rebate of $9,000 and an out-of-pocket cost of $10,000. Also, until the year 2016, there is a federal tax credit of 30 percent of our out-of-pocket cost, so, based on our taxable household income, we expect to also maximize our income tax credit this year for another $3,000 cost reduction. This brings our net after rebate and tax credit cost to $7,000, and based on our expected annual utility service costs over the next few years (average $0.15/kilowatt hours), we can expect to pay for the solar system in less than 7 years. (Note that panels must be SRCC certified for rebates and tax credits. Rebate amounts may vary. Rebates and credits apply to primary and secondary residences.)

Since most areas of the country do not have any local incentives for solar, the 30 percent federal tax incentive would be the only offset to the initial cost. That amounts to a $5,400 reduction, or net cost of $12,600. Areas of the country that are already paying demand charges may see a faster payback. We think in general that the systems installed in the market today are averaging a 10-year payback, as compared to average 20-year paybacks just 5 years ago! Of course, this all depends on making the house as efficient as possible so that the loads are reduced to match a smaller systems production capacity. Although we do not expect that every homeowner will install a solar photovoltaic system, we do hope that, by designing and building a home that optimizes energy efficiency, the owners will see the advantage of adding solar at some point in the future.

The Future?

What about future opportunities? Nanotechnology is developing window films that change seasonally to keep the heat out in the summer and actually absorb it in the winter, and many other building envelope materials are starting to use phase-change technologies. As these developments advance and become more mainstream, we will certain see easier opportunities for net zero energy with even smaller onsite generation systems required.

We'll also probably see a movement in the future to implement performance-based building codes, rather than our current prescriptive codes that define individual efficiencies for various building materials and systems. We should change the way that we look at our buildings. We can adopt a mindset that defines how much energy a home *should* use and then say that is the basis for design and construction, regardless of home size, building components or whatever. You would just have to present an energy model that shows compliance with the acceptable energy factor. In the near future, we expect to see a paradigm shift to an expectation that homes be net zero energy capable and contribute a certain level of quality to our overall quality of life. Won't that be lovely?

We wish all of you much success in your efforts. Remember, it takes a team with expertise to get you there, so recognize the value of each contractor that you bring on board for your project, and make certain that your builder checks their credentials with multiple sources that can verify this is not their first rodeo. You need proven records of achievement. We have seen many successful net zero energy capable homes built in the last few years, so there is some experience out there to be shared.

CHAPTER 12

Net Zero Water

This chapter is dedicated to changing the common perception of where water comes from, how much of it is ours to use and how we can live using only our fair share. Water shortages have become a global concern, with climate change, population growth and pollution placing continued strain on available ground and surface water supplies. Every region of the US has experienced water issues over the past few years, some even outside of drought conditions.

Also across the US, aging water and wastewater infrastructure systems are in desperate need of major renovations and replacements, wasting unacceptable volumes of potable water to leaks, while sprawl demands we use our available funds to build an ever-expanding network of new installations that also will require maintenance over time. It is uncertain that we can ever commit the tremendous amount of resources and funds needed to fix these problems. So before we take any steps in that direction, we should stop and analyze whether the infrastructure systems of the past are the best solution for managing water resources in our future.

The title of this chapter is not a premonition of the future of water on the planet (we hope!), but a visionary goal of what we might achieve some day in terms of living sustainably on our available fresh water supplies. In some areas, current legislation and building codes prohibit some of the strategies that we will discuss here, which is even more reason to bring them to the table. In the future, we will need to remove these unnecessary hurdles so that we can progress toward mainstream water independence.

Most of the strategies to be discussed here are currently being used in areas where public surface water supplies are not available and the cost of drilling for groundwater is prohibitive. These methods provide responsible solutions that can be implemented in urban as well as rural settings, on a large, development-wide scale or in individual applications. People in various communities are making efforts to work through these issues within their particular jurisdictions, so currently we can only caution you to check with your state and local water quality regulators for how these strategies can be applied in your area.

Also, we'd like to note that some topics discussed here may question current policies and practices with regard to individual water rights. We continue to see stopgap measures where entire communities install permanent infrastructure and miles of new water lines to tap remote water resources when their existing local supplies can no longer address demand. As extreme weather events result in demand exceeding supply, they partner with neighboring communities to share available resources. This is just a band-aid that will lead to a bigger problem in the future, as those resources also become overextended.

This chapter's content is based on the premise that water resources belong to everyone in a local or regional geographic area defined by shared water resources. This includes rivers, lakes, aquifers and natural springs, as well as rainfall, that may provide water supplies to vast expanses of land. These areas usually include cities and towns, agricultural developments, ranches and industry. Many of our current water-right laws have upheld the concept that if you can pump it, it is yours to use (right of capture)—the idea that we each have a right to unlimited supply. This is simply not sustainable, especially as our global populations grow and migrate toward certain urban areas that are attracting population growth.

There are instances where agricultural developments have been denied water, so that urban development can continue in unsustainable practices. Ranches have been forced to sell off their stock due to inadequate rains for growing hay or filling ponds. We need to reevaluate and redefine our priorities with regard to two of our most valuable resources—water and food. We must find ways to manage both for the greater sustainability of our ecosystems and our own sustenance.

As our global population continues to grow and more people migrate to dense urban areas, at some point there will be too many straws in the well of local water resources. As climate change continues to move many areas to a more arid climate, periods of extended drought will lead to decreasing natural water supplies. As populations increase and water supplies dwindle, how will we have enough water? Certainly we will need to make some difficult choices.

The Water-Energy Connection

The treatment and pumping processes required for municipal water and wastewater services can be the largest consumer of electric energy for the region that they service, so saving treated water can also significantly impact energy production needs. One of the interesting findings of a study published by the EPA in 2008 was that 50–78 percent of typical residential water use was in non-potable areas.[1] The energy savings alone, to treat water to potable standards, including that required for pumped delivery to the customer, is substantial. Taking these non-potable uses off the table would result in huge savings in energy. This in turn would reduce our need to build more power plants, as well as eliminating any related pollution from those processes.

Additionally, pumping wastewater (blackwater) to sewage treatment facilities requires tremendous energy and water resources in and of itself. Treating wastewater onsite (septic systems) or reducing wastewater, through onsite composting, should be further considered in order to reduce municipal infrastructure and pro-cessing inefficiencies. Not only are we realizing that water is a valu-able resource, we can also recognize that this type of "waste" has value as well.

Where Does the Water Go?

The remainder of this chapter will follow the course of what each of us can do to be part of the solution or at least stop contributing to the problems we have discussed up to this point. We will present possible solutions that each of you can incorporate into your con-struction project to secure a more sustainable water future for your family. We will address those strategies first that provide the biggest impact. To determine what those efforts will be, it is important for us to analyze where we are currently using the most water and to what extent each use requires water treatment. So this section will look at consumption from the biggest piece of the pie down to the crumbs. For those of you who might question the importance of the little things, just remember that, as we make significant achieve-ments in the larger arenas, those little things represent a larger por-tion of what remains to be addressed.

Rethinking Our Landscapes

It's probably no surprise to anyone that 50–70 percent of total household water usage continues to go to the landscape.[2] This is the greatest portion of the non-potable water use described in the 2008 EPA study mentioned above. So reductions here will have the greatest impact. The first issue we must look at is our current landscaping practices. Many new developments have mandated landscape packages, defined by an architectural control commit-tee, neighborhood deed restrictions, or government agencies like the Federal Housing Administration (FHA) or Veterans Adminis-tration (VA). Many are defined for aesthetic consistency, not water efficiency. Although some residents might agree that a common interest in maintaining property values might allow these author-ities to dictate such details, they often lack information on alternatives that could achieve both attractive landscapes and water conservation.

There has been some press coverage lately regarding home-owner association deed restrictions that penalized members who followed mandated water restrictions during drought periods at the expense of their prescribed landscape maintenance requirements. In response, contractors have started to install artificial turf to provide that manicured lawn look. As green building consultants, we have been asked for our opinion on whether that could be considered a green feature, as it requires no watering. We think that there is a trade-off between the water savings and the greenhouse gas emissions resulting from the manufacture of that product, as well as the embodied energy to get it to the site. Artificial turf doesn't do anything to support nature or wildlife. We have to think that there are better alternatives, including xeriscapes, wildscapes and mulch or rock gardens.

Soil Amendments: Adding high volumes of organic soil amendments and continuously improving the quality of the landscape soil will increase its ability to hold moisture, thereby reducing the frequency of supplemental watering. Expanded shale (hadite) is a porous rock that holds 38 percent of its weight in water,[3] releasing it back into the soil as the soil dries out after being wet. When used as a soil amendment, especially in heavy clay soil, this material can extend periods between scheduled watering. Better yet, this amendment only has to be added once since it does not decompose like compost.

Soil amendments should be used in all areas of the landscape, including turf areas. The right soil conditions provide the best environment for any type of plant—grass, flowers, shrubs or trees—to flourish under even the most extreme conditions. Compost and shale not only improve the water-holding capacity of the soil, they also improve drainage and create the right living environment for microorganisms, which feed and support the plants' ability to absorb nutrients from the soil.

Plant Selection: Next, select only native or adaptive plant species for your area's climate, especially those that are able to live off average rainfall. In an arid region, succulents are a good choice, while in a marine climate, choose only water-loving species. The best selection of plants for your area will be the same ones that you see growing in natural environments locally.

The Right Irrigation: It is important to use supplemental irrigation for the first year, in order for the plants to establish healthy root systems. After that, wean them off supplemental irrigation to get them acclimated to natural rainfall patterns. Supplemental irrigation should only be used then in periods of extended drought.

When it comes to making difficult choices, the first should be to eliminate unnecessary outdoor water use. In consideration of this, mechanical irrigation systems should be avoided. Using supplemental irrigation indicates that you have selected plant species that cannot survive naturally in the landscape. Using it on a continual basis will only cause the landscape to become dependent.

If you must install an irrigation system, make sure that you hire a WaterSense-certified installer and that the system includes high-efficiency heads, a rain sensor and a soil moisture sensor to optimize its performance. These systems often pay for themselves in less than three years in water savings and yield a healthier lawn. They measure rainfall, soil type, humidity, solar exposure, wind and soil moisture levels using wireless probes then run all of that data through an algorithm to determine when and how much to water each zone.[4] It is better to run the system to deliver one inch of water once a week than to water half an inch twice per week. Or better yet, skip the automatic settings and just run it manually when it is really needed.

The only exception to this is if you are able to capture water onsite specifically to provide natural irrigation for your landscape. This will be discussed later in this chapter when we look at available water sources. However you source your water, it is important to use it wisely, otherwise it may actually be a waste of both the resource and your efforts. Remember, every drop that you capture is one less drop of runoff that replenishes ground and surface water supplies that are dedicated to other water needs (like farming for our food supply).

New methods of landscape irrigation are also gaining ground. Trench beds utilize mulch as a base material to capture and absorb excess moisture and release it back into the bed, as needed. You can also use straw bales to border raised beds and achieve similar results.

Wicking beds involve encasing the landscape bed in a watertight enclosure, either using raised bed systems or by excavating your landscape bed and lining it with a heavy, impermeable pond liner. Add a 4-inch perforated pipe along the bottom of the enclosure (use flex pipe or a 90-degree connector), bringing the end of the pipe up to the level of the top of the enclosure. Provide an overflow drain hole in the side of the enclosure about the same level as the top of the perforated pipe. Then cover the bottom of the enclosure to a depth above the top of the pipe with clean, small gravel.

Use the drain pipe to fill the bottom of the reservoir with water, up to the top of the gravel. Fill the remainder of the enclosure with a blend of good-quality planting soil and compost. The soil layer

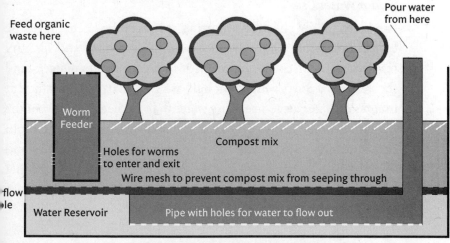

FIGURE 12.1. Wicking bed. Note the added vermiculture feature in this illustration.

should be at least six to eight inches deep. Plant your landscape as you would in a traditional planting bed or use this system for your vegetable garden. The plant roots will search for the water, developing deep roots down into the gravel bed. Add water, as needed, using only the perforated pipe, as this bed captures and retains the water with minimal evaporation or irrigation losses. The overflow pipe will indicate that the proper depth of water has been achieved. This is the most efficient irrigation system that we have ever come across, so it will have the greatest impact in reducing landscape or gardening water requirements.

Minimize Evaporation: Don't forget the mulch. A good one-to-three-inch layer of mulch over any type of landscape bed, including your vegetable garden, will also significantly extend periods between watering. Use mulch from local materials, as it breaks down over time into compost that provides nutrients missing to the native ecosystem, further improving soil quality. Dry-farming methods use mulch as walkways between planting rows, so the mulch captures runoff and, again, wicks moisture back to the soil as it dries out.

Indoor Water Use

Now let's follow the water flowing through the meter into the typical household. Many builders and homeowners still identify residential water conservation with low-flow toilets, faucets and showerheads. Surprisingly enough, we find that while inefficient fixtures do cause us to use more water than we need to, most of the water consumed inside the house is being wasted down the drain at various sinks and showers after you turn the faucet on and stand there waiting for the hot water to arrive. It is most important that we make improvements to reduce that waste before we invest in more efficient fixtures for delivery.

Wasted Down the Drain: If you are building or remodeling, your best opportunity to address this is by incorporating a central core plumbing design with a centralized location of the hot water heater. Other alternatives include adding tankless gas water heaters located near each of your plumbing zones (a very expensive solution), or installing an on-demand recirculation system at the end of the plumbing runs, with controls placed at each wet location along that zone run (see Chapter 3).

Appliances and Fixtures: By following the water through the pipes, we can determine how much of it goes to each use, and we can recognize opportunities to improve each application. The pie chart below from a 1999 American Water Works Association Research Foundation (AWWARF) report, "Residential End Uses of Water Study,"[5] shows that water used to wash clothes and flush toilets represents a significant portion of the indoor consumption. This indicates that over one-third of the water used inside our homes could come from non-potable sources. This is a great opportunity to use untreated rainwater for laundry activities and graywater

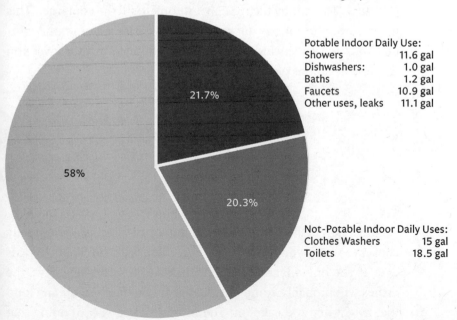

FIGURE 12.2. Domestic water use: clothes washer 21.7%, outdoor use 58%, all other 20.3%. Credit: Adapted from AWWARF.

for flushing toilets, eliminating those completely from the potable water treatment equation.

Once you have addressed efficient delivery of water and water reuse opportunities, it's time to look at fixture efficiency. The most readily available options are high-efficiency toilets (HETs) or dual-flush models. As water resources are further stressed, composting toilets may become more accepted and used in order to reduce the size of this piece of the pie. Current federal limits on faucets are 2.5 gallons per minute, but there are many models on the market now with lower flow rates, or you can install aerators to reduce flow without compromising the feel of the flow necessary to wash your hands. Same for showerheads, which are now available in less than two gallons per minute flow rates.

If you do not know how many gallons of water your fixtures flow, you can put a one-quart container below the faucet and turn it on to its maximum flow rate. Watch the second hand on your watch to determine how long it takes to fill up the container with water. Divide 60 seconds by the time that it took to fill the container. This will give you how many quarts you can fill with water in one minute. Your answer (including its fraction) divided by four gives you the flow in gallons per minute.

> If it takes 5 seconds to fill up a quart container with the faucet turned on all of the way:
>
> One minute = 60 seconds/5 seconds =
> 12 quarts filled in one minute/4 quarts in a gallon =
> 3 gpm flow rate for fixture.

Most new dishwashers do not require pre-rinsing, and to really conserve water, you should only run them with full loads. In this case, if you only fill and run your dishwasher every few days, the dishes will probably come out cleaner if you do rinse off any food that might dry and cake on. The most efficient way of doing that is to fill the sink with as little water as you need to dip each dish in, swipe with a rag and transfer to the dishwasher. Look for new models that are ENERGY STAR rated for both energy and water efficiency. This is also true for clothes washers. Developing technologies will soon support other water-conserving appliances as well.

Changing How You Use Water

Your lifestyle choices and habits can also significantly reduce indoor water use. Showering typically takes much less water than bathing in a tub, especially if the tub is a large soaking or jetted type If you have teenagers living in your household, they might be taking long showers without consciously realizing the water waste involved. Teaching the importance of water conservation is part of a comprehensive sustainable living curriculum and might be best approached by setting an example when children are young and you are actively engaged in their bathing. This includes training in the use of any mechanical controls that reduce water waste, as well as choosing the correct source of water for each activity. If you have installed a graywater system, education on what it should and should not be used for is critical for both water conservation and health safety.

Residential bath and shower controls are now available with flow control valves that allow you to manually pause the water flow during bathing. This is similar to fixtures that have been available for years in the recreational vehicle market. If you've ever been camping and stood below a 5-gallon water bag to take a shower, it's a similar experience. You get wet, shut off the flow, soap up and then turn the flow back on to rinse off. It's surprising how many people can bathe off a 5-gallon container, as compared to 30–50 gallons for the average home bath, or 20 gallons for the average 4-minute home shower.

Similar water savings can be achieved washing your car if you use a water hose with a control head attached. The average person uses over 500 gallons of water annually washing their car![6] With a small bucket and hand-washing sponges and rags, you can wash a car with a gallon or two of water and just enough pressure from a spray attachment to rinse it off. The same is true for power washing your deck. Again, this is a good application for rainwater, no need to use potable water for washing the car or deck! In fact, there's little reason at all to connect outdoor hose bibs (faucets) to a municipal water system, as long as users (including kids!) are well-educated that this is not a drinking water source. Note that water-quality regulations may require that these sources be clearly labeled as not for drinking. In fact, we can reeducate ourselves to expect that all of our outdoor water needs will be met with non-treated rainwater, unless you have an outdoor kitchen or shower, which should be connected to your potable water source.

Water leaks are also a big contributor to wasted water. You can add a little food coloring to your toilet tank to see if it leaches into the bowl, which indicates a water leak. Keep an eye on exterior hose bib faucets, as leaks not only wastewater but can damage the foundation of your home. Paying attention to your water usage, whether examining your bill from the water utility provider or by watching the level line on your rainwater tank, can alert you to possible leaks before they get expensive to deal with and require you to pay for water you did not get to use.

The Future of Water Use: Already, much of the world has ongoing water availability and safety concerns. We believe that, at some point in time, we too will be faced with choosing between outdoor landscape water use and our more basic water needs. We may need to prioritize water uses in order to assure water supplies are available for local food production, dedicating available groundwater and surface water to assuring that we have local sustainable food sources. But at some point, we are going to have to move beyond the low-hanging fruit to delve deeper into how to make more substantial improvements in water conservation. This means looking at not only what we use water for, but also where it comes from and where it goes after we use it. To have ample water resources in the future, we are going to need to recognize a strategy that provides each of us only our fair share.

Using Only What We Can Capture

As society further recognizes the cost of large-scale district or municipal water treatment plants to treat water to potable water standards, as well as the cost of building and maintaining large potable and wastewater infrastructure systems, moving more to localized systems makes more sense. But using a community-wide treated water supply as a sole source of water for landscape purposes is an expensive waste, especially in terms of the energy used to treat and pump it. This cost can rise exponentially if expensive lift stations are required in your area due to topography. Rainwater makes much more sense, and it is better for your plants as well. Additionally, the fluoride and chlorine added to municipal water kills all of the beneficial microorganisms that convert soil nutrients into a form that plants can use.

In a net zero water world, each home would be limited to the amount of water that can be captured, used and reused on its site. Many people in the world subsist only on this model. Think about those populations that live on islands, places like Bermuda and the US Virgin Islands, where homes are designed for rooftop collection and underground cistern storage. Relying only on their ability to capture and store rainwater and water reuse has provided all their water needs since such places were inhabited.

For the purposes of this chapter, we will also use the model where all our water is captured through rainwater collection. If our goal is a sustainable water source, what other means is as reliable as this? Rainwater can provide non-potable use, as well as indoor potable water use. Rainwater is a renewable resource and an excellent source of high-quality water. "Captured onsite" does not mean the ability to drill a well into shared water supplies and have unlimited access to those supplies. Ground and surface water resources must be set aside as dedicated to ecosystems, wildlife, agricultural needs and emergency and fire defense. We are going to limit those enough when we all start capturing and using as much of the water that falls on our sites as we can.

To achieve water independence, we must incorporate numerous methods of capture and provide separate storage facilities for various uses, each defined by quantity and the type of sanitation treatment required. Some water could be captured and stored within the land itself. Other capture will require some type of container or cistern for storage, but no type of treatment for its use. And, finally, water for potable purposes would require separate storage in a food-grade cistern, with an onsite treatment system. Beyond the various initial means of capture and use, reuse of water will require additional systems for storage and treatment. All of these activities require significant thought with regard to separation of plumbing pipes to protect water quality and safety.

If we look closely, we can see that this is a viable basis for developing sustainable water conservation practices. For some dense developments, this may mean that a community-wide system is in place to capture water on all rooftops for centralized storage and distribution. We currently are seeing this type of design applied to renewable energy production, utilizing the rooftops of commercial buildings to house large solar arrays to provide electric power to surrounding neighborhoods. At some point, those same rooftops may collect rainwater to be stored in large reservoirs beneath the foundations of those buildings for the same purpose of supplying neighborhood needs. Only when these types of developments share community agricultural garden space would other water resources, including regional groundwater and surface water supplies, be made available.

How Much Water Is Available
Rainwater Collection

To achieve net zero water, we need to first determine the volume of water we are able to capture from our roof area and store, and then how much of that can be used for each application necessary for us to sustain our site. Determining how much rainfall you can capture depends upon how large a roof area you have and how accessible that surface is in terms of installing a gutter, downspout and piping system to carry your capture to storage tanks. The storage capacity you will need is determined by looking at the rainfall patterns for your area and assessing the size of containers needed to manage enough water to meet your family's needs over time. This is usually calculated on an annual basis, considering that we may get our rain seasonally yet need to store enough water to meet our need during the non-rainy periods of the year. As an example, Austin, Texas, and Portland, Oregon, get about the same amount of annual rainfall, but Portland gets theirs in frequent mild showers, while Austin's comes in a few severe flashflood events with little rain in between. Austin families must have more storage capacity to be able to last longer between replenishing rain events.

You can collect about 0.6 gallons of water per square foot of roof area during a one-inch rain, or 600 gallons per inch of rain for every 1,000 square feet of collection area. Based on your available roof collection area and the configuration of your gutter system to capture and collect roof runoff, you can determine how many gallons of rain you can expect to collect from each one-inch rainfall. Next you will need to calculate how many gallons your family uses on a daily basis and then think about how long the average period is between measurable rainfall events in the various seasons where you live. This will help you determine how much storage capacity you will need. Of course, the higher your water needs, the larger the collection area and storage container you will need.

You'll need to store enough supply to meet your daily needs in your lowest rainfall season until the next rain, unless you plan to purchase water to keep your tanks replenished. Purchasing water this way (one truckload at a time) can be expensive and adds significant embodied energy, and it doesn't really meet our goal of net zero water, so it should only be done in periods of unexpected prolonged drought. Also, depending on the water source, it probably is not going to be the same quality of water that you will get from pure rainwater.

As mentioned earlier in this chapter, it is necessary to keep separate storage units to meet your different needs. Just as it is a waste

ROOF SIZE IN SQUARE FEET	RAINFALL IN INCHES								
	20	24	28	32	36	40	44	48	52
1000	11236	13483	15730	17978	20225	22472	24719	26966	29214
1100	12360	14832	17303	19775	22247	24719	27191	29663	32135
1200	13483	16180	18876	21573	24270	26966	29663	32360	35056
1300	14607	17528	20450	23371	26292	29214	32135	35056	37978
1400	15730	18876	22023	25169	28315	31461	34607	37753	40899
1500	16854	20225	23596	26966	30337	33708	37079	40450	43820
1600	17978	21573	25169	28764	32360	35955	39551	43146	46742
1700	19101	22921	26742	30562	34382	38202	42023	45843	49663
1800	20225	24270	28315	32360	36405	40450	44495	48540	52584
1900	21348	25618	29888	34157	38427	42697	46966	51236	55506
2000	22472	26966	31461	35955	40450	44944	49438	53933	58427
2100	23596	28315	33034	37753	42472	47191	51910	56629	61349
2200	24719	29663	34607	39551	44495	49438	54382	59326	64270
2300	25843	31011	36180	41348	46517	51686	56854	62023	67191
2400	26966	32360	37753	43146	48540	53933	59326	64719	70113
2500	28090	33708	39326	44944	50562	56180	61798	67416	73034

Annual Rainfall Yield in Gallons for Various Roof Sizes and Rainfall Amounts

FIGURE 12.3. Rainwater harvesting yields.
Credit: Sustainable Sources.

of energy for water utilities to treat supplies for irrigation, it is also a waste of resources for you to use your filters for treating rainwater for potable use for water you will use in your landscape. You should separate your calculations for each end use to determine the volume of storage you will need for each and the various piping configurations to get the water to where it will be treated for use or used without treatment, as necessary.

For indoor potable water, you must install a filtration system and complete disinfection by using ultraviolet light or ozone treatment. For our own health and to protect the environment, the treatment method should be able to provide safe water without the use

Sample Water Balance Calculations for Dallas, Texas
(Using Average Rainfall and a 2,500-square-foot collection surface)

Month	A. Water demand	B. Irrigation demand (watering by hose or bucket)	C. Total demand (gallons)	D. Average rainfall (inches)	E. Rainfall collected (gallons)	F. End-of-month storage (1,000 gal. to start)
January	3,000	0	3,000	1.97	2,596	595
February	3,000	0	3,000	2.40	3,162	757
March	3,000	150	3,150	2.91	3,834	1,441
April	3,000	150	3,150	3.81	5,020	3,311
May	3,000	150	3,150	5.01	6,601	6,762
June	3,000	150	3,150	3.12	4,111	7,723
July	3,000	150	3,150	2.04	2,688	7,261
August	3,000	150	3,150	2.07	2,727	6,838
September	3,000	150	3,150	2.67	3,518	7,206
October	3,000	150	3,150	3.76	4,954	9,010
November	3,000	0	3,000	2.70	3,557	9,567
December	3,000	0	3,000	2.64	3,478	10,000*

* Note that there were 44 gallons of overflow in December in this example. A 10,000-gallon cistern appears to be appropriate under the given assumptions.

FIGURE 12.4. Water storage requirements.
Credit: Texas Water Development Board.

of chemical treatment. For non-potable water use, if you can reduce your needs to a minimum, it would not be cost-effective to separate storage.

Water Reuse

Water reuse may be the last frontier of water conservation strategies. It requires dual plumbing of the home both for separating potable and non-potable delivery and for separating graywater and blackwater discharge. Beyond those secondary uses or treatments, even more additional plumbing is required for different tertiary applications.

Graywater

Graywater reuse is the first layer of this effort. Acceptable graywater sources may vary according to local jurisdictions, but generally accepted sources are lavatory faucets, showers and tubs and the clothes washer wastewater. Some authorities also allow kitchen sink wastewater (except that from garbage disposals), and others prohibit water from bathing activities. You should check with your local regulatory agencies before plumbing for this type of system.

In some cases, graywater has been plumbed from bathroom lavatory sinks to be used again to flush toilets. It can also be carried directly out into the landscape by plumbing that discharges it separately from the blackwater discharge lines from kitchen sinks, dishwashers and toilets that go into wastewater treatment. However much graywater is sent to the landscape may not be enough to reach desired root zone depth, possibly even encouraging shallow root development that exasperates water stress, so it may only represent part of the equation for meeting our outdoor landscape water needs. But as we move away from dedicating water for landscape use, it may be sufficient for what remains. There may be regulations that limit the holding period for graywater storage, so, again, check with local regulators before designing your reuse system.

Studies[7] can provide us with research as to the typical amount of water available from these sources, based on family size and lifestyle patterns. You should make adjustments to any published calculations if your family's activities include showering at a gym or children leaving for boarding school for months at a time. It is also important to note the age of any study research references. As faucets and washing machines get more efficient at using water, the amount left to contribute to our graywater supply may be greatly diminished from published averages. This is especially true of newer clothes washers that only use four gallons per cycle and motion-sensor predetermined-flow lavatory fixtures, which should be on your want list to reduce overall water needs.

Condensate

The reuse category also introduces new sources of water for available use.[8] Capturing condensate from air conditioning systems can provide significant sources of non-potable water in areas of the country with high humidity levels and hot summer temperatures. Additionally, condensate can supplement untreated rainwater for outdoor use during extended drought periods, when rainwater stores might otherwise be exhausted. In green mixed-use developments, condensate is available not only from our own residential air conditioning systems, but also from the excess generated by commercial systems, especially those on large office buildings, hospitals and industry located within the same development. The same is true of chilled water from large commercial chillers, which can be distributed both as a means of providing radiant cooling and as an irrigation water source.

Just like other forms of gray water, condensate from air conditioning systems can be combined with untreated rainwater, used, then reused as graywater or treated for reuse. Although condensate quality is about equal to distilled water, exposure to contaminants in systems and storage containers does not allow for its use as potable water. However, it serves well as a supplement to irrigation water, as well as for clothes washing and flushing toilets inside the home.

Care must be taken when reusing water from many commercial HVAC chillers (different than condensate) since it is often treated with chemicals to prevent corrosion or scaling that could be toxic. This requires community infrastructure to support piping it to the locations where it can be used. It also may encompass various types of water treatments, recapture methods for repeated reuse and storage facility management. Just as our solid waste is now being viewed as a recyclable resource, we need to recognize that what was once considered wastewater is merely water that requires additional treatment for reuse.

Stormwater Runoff

Capturing stormwater runoff on the site can provide another source of irrigation water; it might even be used for some indoor non-potable water applications. The goal is to manage water on our own site using the most energy-efficient methods possible, especially if we also have a net zero energy goal. Remember our basic principle from Chapter 1: water runs downhill. We will use much less energy moving water from our rainwater cisterns to our point of use if that direction is downhill. The same goes for removing wastewater after use: plumbing should flow downhill to the reuse or treatment system. Keeping this downward flow going as much as possible through all of our processes means using the least amount of energy for those processes.

For landscape use, water can be diverted through a series of berms and swales or use an underground French drain system. This water is directed to landscape beds or stored in retention ponds and pumped out as needed for various uses. Other permanent erosion control features can also help to manage and keep water on the site.

Permaculture methods[9] can capture runoff on the high side of trees, allowing it to infiltrate at these points, feeding water down to the tree roots by natural flow and subsurface distribution. This technique encourages the use of materials available onsite, which can include brush, mulch or downed tree trunks and branches, to slow water down as it moves through its natural course downhill. In landscapes with a significant slope, retaining walls should be built to terrace the area, reducing erosion of native soils and runoff and improving infiltration. The more we can use these types of features to keep natural rainfall on the site, especially soaking in to promote deep root growth, the less supplemental irrigation we need to provide. In this way, literally all the rain that falls on your site that does not infiltrate immediately into the soil can still be used.

Onsite Treatment

We need to develop onsite methods to treat water to varying degrees, based on the level of water quality needed for each different use. This includes the ability to treat "water waste" onsite for use again and again, if possible, in a closed-loop system. We must also treat any wastewater to be discharged from the site so as not to cause damage to our soils, neighboring groundwater resources or ecosystems.

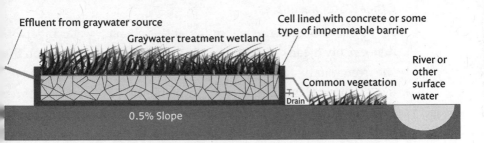

FIGURE 12.5. A typical subsurface flow graywater
wetland system.

If you have a large household that generates a large graywater
volume, beyond your ornamental landscape needs, you should
consider onsite treatment that would purify the water for edible
food crop use. Biofilters[10] can remove many organic substances that
contaminate water during bathing and clothes washing. Of course,
the use of biodegradable cleaning products is necessary to prevent
chemical contamination that cannot be treated onsite. At a single
household or on a community scale, constructed wetlands are able
to process and return graywater to a usable state for a very low cost,
removing pathogens, bacteria and non-biodegradable toxins. And
the beauty of this system (other than the fact that it can be a beautiful addition to the landscape) is that it can treat the same water
repeatedly for reuse.

Of course, various types of septic systems can be used to treat
and manage blackwater and fecal waste, but those are dead-end
systems. Composting toilets, however, provide the benefit of not
requiring water to flush waste through like conventional sewage
piping does. Some models are now available with urine separation,
which provides composting opportunities for human feces without
the need for blackwater treatment at all. Urine is naturally sterile
and can be used immediately as a valuable fertilizer.[11] Human waste
can be processed as manure for uses other than food production.
Care must be taken whenever using human feces to ensure that it is
thoroughly composted and raised to the necessary temperature to
make it safe to handle.

Other Water Resources

As we mentioned at the beginning of this chapter, water independence really means that we are able to meet all of our needs on the site from water—rain and condensate—that we can capture and re-use. Of the vast amounts of water available on the planet, only three percent is freshwater, and most of that is ice. Less than one percent is readily available for use by humans.[12] So many are considering and investigating desalinization of the vast supplies of water in the world's oceans.

In Chapter 14, we will look at the issues contributing to and caused by carbon emissions on our planet. One of those that we will mention relates to the amount of carbon that is sequestered in those vast ocean areas. The ocean stores approximately one-third of the carbon sequestered on Earth each year,[13] but as carbon emissions grow, this is causing ocean acidification, threatening this essential ecosystem. The increase in the acidity of the ocean is contributing to the decline of the coral reef systems around the world due to the lack of calcium carbonate available to build and maintain their structure. Coupled with overfishing and increased pollution activities, as well as warming of ocean temperatures, fish populations worldwide, a major source of food for many cultures, are in decline. Many are already facing possible extinction.[14]

Many desalinization methods[15] dispose of the removed salt, or brine, in ways that can cause significant contamination to our soils, groundwater and surface water supplies, or further damage our oceans. Treating one water source to contaminate another (or others!) just makes no sense at all. Coupled with the tremendous amounts of energy and infrastructure needed to manage the treatment and distribution processes, and the greenhouse gas emissions resulting from the operations, the result is not unlike the unsustainable water systems we already have in place.

Case Studies

There are plenty of studies available on the numerous examples of people who are, for one reason or another, living off the water grid.

These range from areas without municipal water connections and unreliable groundwater supplies to those living in remote locations, like mountainous areas where rainwater catchment is a less expensive investment than drilling a well for groundwater. It's not a matter of whether or not this is something that we can accomplish, it's a matter of when it will become mainstream.

Although there has been research into applications of the strategies discussed in this chapter, we are not aware of any particular instances that incorporate them into a net zero water scenario that fits our model. So, we've created a vision of a closed-loop system that would accomplish this goal.

- Roof designed to capture sufficient rainwater: storage capacity based on annual precipitation and indoor potable water need of family with backup capacity for fire protection or in case of extended drought, would supplement water needs for food production
- Site designed to capture and use remaining stormwater
- Graywater system installed to reuse water from laundry, lavatories and showers; discharged to biofilter pond
- Edible landscape installed in wicking gardens built using expanded shale and compost, with a good mulch layer maintained
- Fruit and nut tree soil amended with expanded shale and compost, with a good mulch layer maintained
- Remaining landscape is wildscape (natural habitat restoration), reusing native soil amended with compost
- Composting toilets installed so no human waste enters wastewater system
- Blackwater from kitchen sink directed to biofilter
- Permaculture techniques used in landscape to direct water to retaining ponds
- Condensate lines drain HVAC water into dedicated storage tank with water from biofilter
- All indoor potable water supplied by rainwater filtration system
- All laundry water and garden water sourced from biofilter/condensate storage tank

- Stormwater retention pond(s) used to irrigate edible landscape garden and trees

Again, make sure any of the strategies you choose for your project are allowed by local and state laws. If nothing else, many of these can be incorporated into your outdoor landscape water management plan. Permaculture techniques have long been used to manage stormwater runoff as a source of landscape irrigation. Improved gardening methods can significantly reduce the amount of water needed. These are steps in the right direction that we can all benefit from.

FIGURE 12.6. Living Building Challenge water petal.
Credit: International Living Future Institute.

How Low Can You Go?

The issues discussed in this book are not specific to the United States or North America. Many areas of the world have much worse conditions, and others have centuries of experience we can all learn from. To participate in this worldwide sharing, join and contribute to the Living Building Challenge (LBC). To learn more, visit living future.org/lbc.

Living Building Challenge Water Petal Imperatives

05 Net Zero Water

One hundred percent of the project's water needs must be supplied by captured precipitation or other natural closed-loop water systems that account for downstream ecosystem impacts, or by recycling used project water. Water must be appropriately purified without the use of chemicals.

06 Ecological Water Flow

One hundred percent of stormwater and used project water discharge must be managed onsite to feed the project's internal water demands or released onto adjacent sites for management through acceptable natural time-scale surface flow, groundwater recharge, agricultural use or adjacent property needs.

"Living Building Challenge™" and related logo is a trademark owned by the International Living Future Institute™ and is used with permission.

CHAPTER 13

Zero Waste

Waste is a by-product of nearly every activity on Earth, representing tremendous "waste" of a variety of resources, everything from natural raw materials, to the energy and water required to produce products. Our neglect in controlling or eliminating waste adds unnecessary cost to products providing no useful benefit. When we consider the costs of waste disposal and the related impact on the environment, our total cost of ownership of products is greatly increased. But as the old saying goes, "One man's waste is another man's treasure."

Getting to zero waste requires employing two strategies: opportunities to reuse or repurpose materials at the end of their useful life and avoiding the acquisition of materials that cannot be reused. Many efforts are being made in this regard on many fronts, but we will limit our discussion here to only those that we have the ability to influence in our daily lives. Some of these require us to use our purchasing power to influence the manufacturing marketplace, while others depend upon the choices we make to shift the disposal streams.

Waste Beyond the Household

Many municipalities are working toward zero waste plans for their communities.[1] This process requires thorough examination of the composition of the trash that gets dumped into landfills. It has to be traced back to its various sources, and alternatives to disposal researched for each source for each type of material. All of the various contributors, from office buildings to restaurants and various other types of business and industrial activities, to apartment complexes and private homes, each represent their own set of challenges and require uniquely different approaches. But the lessons learned can be shared and offer benefit across market segments. This chapter examines how to embrace the right strategies so that you do not invest in materials and products that will have little reuse value or may even represent liabilities in the future.

To start, we'll briefly view opportunities outside of the household arena so that we can realize how those affect our lives and our ability to reduce our own waste. Let's dig through a few dumpsters from commercial, industrial and retail establishments and see what kinds of waste are found and what else might be done to divert them to some other purpose. Since all of these facilities are workplaces, all the dumpsters will contain typical lunch and break room waste, including aluminum cans, glass bottles, paper, cardboard and plastic products and, of course, food waste. Many businesses already furnish employees with recycling opportunities for paper, plastic, glass and aluminum. Many municipal trash services now have separate facilities for managing and reselling those reusable materials. This has, in many instances, proven to be a new revenue stream for cities, making their efforts more worthwhile. This has also resulted in many of the same recycling opportunities being offered to residential waste customers, providing us with an easy means of recycling these same products from our own homes.

Food and Waste

Expanding the recycling opportunities for food-related waste is on the agenda of many zero waste plans.[2] New compostable food packaging and single-use eating and drinking containers and utensils are available, and over time, should replace those that are not. This new market has spilled over to the food packaging products available for purchase in our local grocery markets.

But the biggest opportunity here has to do with the food scraps themselves. The decomposition of food and other organic waste produces methane emissions (a potent greenhouse gas) from landfill facilities. This creates a problem for the atmosphere, and is a waste of good organic materials. Efforts are being made to provide managed diversion of this waste stream from business establishments to a facility that can use it, along with grass clippings, leaves and other decaying organic matter, to produce quality compost to be used to rebuild depleted soils. This effort will significantly impact the restaurant and fresh produce retail industries, capturing vast volumes of organic waste and redirecting it to produce organic fertilizer to grow more food crops. Backyard composting efforts are being piloted in many cities, again providing us with means to divert our own organic waste streams.

Commercial Waste

The first step in reducing commercial waste is to avoid generating it to begin with.[3] Many types of communications that were historically in printed form are now available electronically, so that method should be utilized as much as possible to reduce the volume of paper waste created.[4] Many office buildings are installing hand dryers, to replace paper products used in restrooms and lunchrooms. Most commercial, office and industrial paper is shredded, often by an independent service provider that is also tasked with its disposal. Many office buildings and commercial establishments now manage recycling dumpsters, especially for cardboard shipping boxes. Again this has helped to create the market for household recycling service pickup of both paper and cardboard products. Those that cannot be recycled should be diverted to our backyard compost bins.

Packing materials are also reusable or recyclable, and efforts to divert them to businesses that specialize in providing shipping services are well appreciated. There is still a need to find reuse or recycling opportunities for plastic package wrapping, although alternative materials are now available that are compostable, so eventually these will also go into the organic recycling stream. These are currently available from retail outlets online, but you should start seeing them soon on your market shelves.

Manufacturers of office equipment supplies typically take back spent cartridges and refill them for resale. Office machines and electronics are recyclable, with many manufacturers now taking back, refurbishing and reselling products. These opportunities exist for home office equipment and supplies as well. We should be looking for manufacturers making efforts to redesign electronics to make it easier to upgrade to newer technologies without having to completely replace systems.

Industrial Waste

As mentioned in Chapter 5, more manufacturers are adopting sustainable practices, including improved management of their facilities. Some have achieved certification for environmental stewardship, and have made efforts to reduce and recycle waste generated by their processes.[5] This may include recycling raw materials into new products, reuse of water resources onsite or use of methane gas resulting from operations as a fuel source, to generate electricity for operations. We should support those efforts by watching for the ISO 14001 label on packaging of products.

More and more manufacturers are also analyzing their product lines and making efforts toward green alternatives. This includes a review of raw materials used, recognizing the impacts on workers who handle and install the materials in the field and looking to the future in terms of life cycle impacts. As discussed in Chapter 4, many manufacturers now realize that their business practices and products are under public scrutiny and that these factors affect their long-term profitability and, therefore, survival in the marketplace. Many have also seen diminishing natural resource availability impacting their product pricing, lowering market share. These factors support efforts to reduce waste and conserve resources to keep products affordable. We should look for products made by companies who are making stewardship efforts known (e.g., free trade or FSC-certified), that are made with recycled content and that provide opportunities for us to recycle materials back so that we can support their efforts.

Hazardous Waste

Efforts made since the 1970s by the Environmental Protection Agency have resulted in the reduction and better management of toxic waste disposal sites.[6] We have continued to improve our own efforts to properly dispose of the consumer products that we purchase that fall into this category, as well as to contribute to the effort to clean up historic disposal sites (i.e., brownfields) for redevelopment.

However, we have at the same time developed new technologies that may lead to future problems from new sources. As a society, we continue to tolerate new unproven methods, especially if they provide us with cheaper resources that are in mass demand. As mentioned in Chapter 7, many thousands of chemical substances have been created over the last hundred or so years but only a small percentage of these have ever even been tested for toxicity. We have no idea what impact they have on the health of our families or what disposal of the products made from these chemicals does to the environment. Until we demand more accountability on the part of manufacturers, we continue to be exposed to products both through using them and in their impact on the environment through unregulated disposal. We are all subject to finding out the consequences too late. This is why it is increasingly important that we investigate the chemical composition of all the products that we purchase, making wiser purchasing and disposal decisions.

Construction and Demolition Waste

With regard to construction and demolition waste, getting to zero will require completely rethinking a building's design and construction.[7] Design will have to incorporate the use of open flex space, intended to be reconfigured over time, as the occupants' needs change. It will incorporate strategies focused on designing to template specifications and manufacturing in a production setting.

Any waste that can be recycled will be used in the original product manufacturing facilities. Jobsite waste will be sent back to facilities for reuse in making new products.[8] Small wood scraps can be processed into engineered wood products, and gypsum drywall scraps can be ground back into new panel product. And all products will be made and installed in such a manner as to enable their deconstruction and reuse or repurposing at the end of their current life cycle. You can start that effort now, designing for flex space and changing needs over time, and the eventual deconstruction and reuse of many of the building components of your home.

Waste Prevention

In order to achieve zero waste, we must analyze which waste streams can be avoided to begin with. Most products have a useful life, so the question is whether the materials they are made from can be deconstructed and reused. This should be a consideration in our purchasing decisions, so that we are not supporting the manufacture of products that have single-use life cycles, especially those that cannot be recycled or repurposed.

We should just stop manufacturing all products made using toxic materials, but what are the chances of that happening? So, in the meantime, we should only purchase non-toxic alternatives. We should also support legislation[9] requiring that all products be manufactured in such a way that they can be returned to any retail outlet that sells them, to be sent back to the distributor, then to the manufacturer by the same truck that brings more goods to market, if they cannot otherwise be reused or recycled. Manufacturers should be focused on refurbishing and recycling materials taken back into new products to save resources, reduce the waste stream and keep products affordable.

Eventually, we should see more instances of retailers offering recycling for the products that they sell. This means that we must press for government policies that address and remove any legal or market obstacles that currently prevent this activity, as well as create incentives to promote it. Additionally, we need to impose fines and hefty landfill fee structures to discourage the easiest and cheapest means of disposal of any materials that can be repurposed. Landfills should be the choice of last resort.

Daily Needs

Many products have external packaging that ends up headed for the landfill as soon as you purchase it.[10] Some packaging is not currently recyclable or compostable, so we should be moving toward packaging that fits either of those criteria or reusable packaging (like reusable vegetable or market bags). Stay away from products that offer no alternative to waste, like plastic foam food and beverage containers. More products are available now in 100 percent recyclable packaging, made primarily from cardboard and plastic. With the new single-stream management services, recycling is as easy as throwing the package into a dedicated recycling receptacle.

Consumer behavior modification can also significantly reduce the amount of waste generated for disposal. This starts by adopting smart buying habits. It also involves how we value a product's durability and useful life and what attempts we are willing to make to divert products that we no longer need from the waste receptacle.

Rather than buying trendy items, save your money and our resources by purchasing classic styles and quantities that you will use over the long term rather than just today. Shop around, so that you make sure that, when you do buy something, you get the best durability and life expectancy available for that product. And make an effort to stop impulse shopping, buying things you don't need, since those products waste natural resources that could be saved for better use. Think about how long you are going to need it to last and be willing to spend a little more to get the one that will not wear out so that you will never have to think about buying another one. When planning a major project, try to think about how its uses might change over time, adjusting the design to accommodate those alternatives.

When you do make purchases, only buy as much as you need. Although that usually means buying less, it might mean buying more if it's an item that you buy repeatedly. For products that you buy over and over again, like toilet paper, packaged beverages or your favorite perfume, buy the largest package you can. Buy these products in larger family packs and split them up into single-use proportions rather than buying individual servings. It will save waste from both the packaging and the product itself when it is used up, plus save you money since larger packages are more economical. Even glass and plastic bottles take a lot of energy to transport and process for recycling, so buy larger jugs and reuse a washable glass for smaller servings. And always carry your own reusable shopping bags.

Non-Toxic Choices

Buy water-based finish products, such as paints, that have less of an environmental impact. If possible, only buy and use natural cleaning and personal care products, organic fertilizers and pesticides and other health and household products. When disposing of any chemical products (paints, varnishes, stains, solvents and automotive oil), be sure to take them to a hazardous waste disposal facility (check with your city or county) so that they are properly handled.

Waste Diversion

Recycle

Following waste prevention, diversion of waste from the landfill is the next biggest factor in waste reduction. This begins by selecting products and materials that can be reused or recycled and then making that effort when the product has ended its useful life. This means actually placing the item in a recycle bin or taking it to a recycling center. Check into local service companies that offer curbside recycling or drop-off locations for different types of reusable materials. Typically, cardboard and paper, plastics and metals (including aluminum cans) are easily recyclable.[11]

Many reusable items can still be useful to someone else, so they should be resold or donated to charity. Many are already participating in this activity, with organizations like Goodwill and the Salvation Army dependent on this to provide jobs and raise funds for other activities. Salvage has now gone mainstream on websites like Craigslist.com or Freecycle.org. Check availability in your area for other recycling or upcycling opportunities, as well. Upcycling is the process where a waste material is used to make a different material than that of its origin. A good example is old tires recycled to make asphalt paving. So before you take your old tires to the landfill, check with the dealer you are purchasing your new tires from to see if they will take your old ones for recycling.

Recycling is also a growing stream of business income within the landfill industry. Landfill operations are attempting, within their own operational processes, to divert reusable and recyclable materials away from those big holes in the ground and turn them into profits.[12] They are finding that these efforts are also a means for making their valuable land area last longer.

Composting

Contrary to common belief, composting is an easy process that requires little effort, has big rewards and, if done properly, does not create a "garbage" odor to contaminate your outdoor living space. Even if you live in an apartment, you can practice vermiculture by keeping a worm bin under your kitchen sink or in a pantry. Red wiggler worms[13] literally eat all of your kitchen scraps, except meat, coffee grounds and citrus.

If you have a yard or even more space, permaculture composting is based on four ingredients: green waste, brown waste, organic soil and water. Green waste takes care of your kitchen scraps, grass clippings, yard weeds and landscape trimmings. You can add meat and dairy if you would like, but these ingredients typically attract rodents and other wildlife looking for food and can decay into a smelly mess, so we recommend staying away from them.

Brown waste is made up of dead (brown) grass, fallen leaves from trees and small woody twigs. Larger wood will eventually decompose, too, but not as fast as you would want your compost pile to decay, so best to avoid it. However, you can line the underside of your compost pile with large branches and limbs—this will help to allow air circulation, which speeds composting. Brown waste also includes some household waste, like single-use paper products and compostable plastic food wrappers. You can add shredded cardboard, newspaper (not the shiny advertisements, though), and office paper to your compost. Soil is just that, good-quality living soil, rich in microorganisms that actually feed on the decaying materials and create compost from their waste.

The key to this method is to pile the three types of waste in approximately three- to four-inch-thick layers inside a designated bin area. Note that the bin should be at least three feet wide by three feet long by three feet deep. Do not make the pile so large that it cannot get air circulation within. Continue until the pile is at least as tall as it is wide and keep it moist.

When the pile is tall enough, start another one and let the original sit for six months. After six months, turn the whole pile over with a garden fork so that what was on the top of the pile is now on the bottom and what was on the bottom is now on the top. Keep it moist and wait another six months, and voilà, you will have made beautiful compost.

Composting yard waste should be part of every home landscaping plan. If you do not have a garden, compost can be used on landscaping or at your shared community garden. Compost is the key to great soil, providing all of the organic matter and microorganisms that are needed to grow healthy plants for your gardening endeavors. It can also be used on grass areas of your yard to increase their drought, disease and pest tolerance.

The Key to Making Changes in Your Life

We have found that the most successful way to change is to take one bad habit at a time and repeat a new behavior until it becomes the new habit. Then adopt another one through the same process. By the way, behavioral scientists tell us that it only takes about three or four weeks to form a new habit.[14]

You might be surprised how much you can reduce your waste just by increasing your awareness of what you buy and what goes into the trashcan. It will also help if you place a dedicated recycling bin and compost pail right next to the garbage can in the kitchen pantry, and if you make the effort to purchase more sustainable products and investigate your other opportunities to recycle or donate.

We have made these efforts ourselves and find they have become our norm now. We hope next to stop purchasing throwaway paper products, except for toilet paper, which can be thought of as a compost item (since this goes into its own composting stream). We have begun to think about our own zero waste plan. How about you?

CHAPTER 14

Zero Your Carbon Footprint

To fully understand the total cost of our housing, we need to consider the impact that it has on the environment. One of the most recognized environmental concerns related to human activity is that of our carbon footprint. This is defined as the sum total of greenhouse gas emissions associated with all of our activities. This chapter will look at how the decisions that we make regarding housing choices affect our individual carbon footprint. Our goal is to reduce those emissions, getting as close to zero carbon as possible, both during construction and over the lifetime of that housing.

Carbon emissions directly influence changes in climate. The carbon produced by human activity has resulted in changes in natural systems, contributing to global warming and causing more severe weather patterns. One of the impacts of these changes has been glacier melt, resulting in rising sea levels. Based on the Intergovernmental Panel on Climate Change (IPCC) AR4 Synthesis Report,[1] in order to avoid sea level rise above one meter, we need to stop increasing and make attempts to stabilize this concentration of greenhouse gas at 450 parts per million (ppm) in the Earth's atmosphere. Sea level rise in itself will devastate a significant portion of the world's population that live on or near coastal areas, as those areas are lost to the sea. But there are many other consequences of this phenomenon. The price that we all ultimately pay will manifest itself in many ways, not the least of which is the mounting debt to finance rebuilding homes as climate change-related disasters become more frequent.

As it turns out, there are a number of ways that our decisions in housing are directly related to our carbon footprint. To fully understand what this means, we must analyze all of the individual carbon footprints of each and every material and service used in every phase of the home's life cycle (see discussion on Life Cycle Assessment in Chapter 4). For the purposes of "getting to zero," we can only imagine what it might be like to build a home that is carbon neutral (i.e., does not contribute more carbon than it sequesters over its lifespan). Studies[2] have shown that 80–90 percent of the carbon related to buildings in cold climates is directly related to operations, while the remaining 10–20 percent is embodied energy from construction materials and transportation. Most of the references that you will see touting "zero carbon," "low carbon," or "carbon neutral" refer only to the operational portion of the carbon pie. In other words, these claims are made solely on the basis of creating buildings that produce as much energy as they consume—net zero energy homes. Embodied energy is not taken into account.

Although net zero energy is one of the most important aspects of carbon neutrality, it does not define a zero carbon home. To do that, we need to reduce the carbon associated with its construction, in addition to that resulting from its operations over its lifespan and end-of-life disposition. Let's put on our "carbon neutral" glasses and take a look at how green building methods support doing that. We hope to provide you with insights to increase your awareness of what activities contribute to your carbon footprint and what choices you can make to reduce it, possibly even making your own attempt at zero.

Revisiting Section One from a Carbon Perspective

We will work our way back through the principles presented in this book for greening your home to look at how each impacts its carbon footprint. How much carbon will be released into the atmosphere by each of your activities will depend upon a number of factors, including which of the strategies we've recommended you are able to use. Many of you will know almost instinctively how some of these strategies support this goal, but may find that other, more abstract, activities have a much greater impact than you could have imagined, and you may have not given them a second thought until now. So, let's take a look at what choices we can make, throughout our Ten Steps in Section One of this book, toward the goal of zero carbon.

Location

The primary factors associated with carbon from the operations cycle of the home include not only its energy and water use, but also those associated with how and where you live.

If your location provides walkable or bicycle corridors to shopping, business and recreational amenities, you can significantly limit your ongoing contribution of greenhouse gases by limiting the use of your automobile. If the location also offers mass transit options, using these for your primary transportation to and from your workplace and other urban travel can make an even bigger impact.

Of course, selecting an urban site will also reduce your life cycle carbon contributions related to sprawl and infrastructure investments. Infill sites with existing community infrastructure reduce all of the carbon associated with new roads and streets, utility lines, schools, playgrounds and public service and private support facilities that must also be constructed and maintained to service new extended communities.[3] Of course, there is also the carbon released from the disturbance of native soils and grasslands for the development to take place. Not to mention the carbon component to building, operating and maintaining any type of new mass transit structure to service sprawl developments. These choices also determine how far goods and services must move to service our continuing wants and needs, and therefore, have a direct correlation to the associated greenhouse gas emissions from the commercial transportation industry.

In order to minimize our carbon emissions related to the construction and maintenance requirements of our home, performing a risk assessment and incorporating strategies into our construction to mitigate damage from risk events can save tremendous amounts of restoration efforts over time, each of which has associated carbon impacts. Selecting a site that provides us the opportunities to incorporate passive solar, passive ventilation and natural drainage patterns means that we can reduce our need for supplemental mechanical systems to maintain our comfort and preserve our landscape.

Grassland ecosystems store almost as much carbon as tropical forests, and soils stockpile more carbon than vegetation and the atmosphere combined.[4] Limiting the area of the site that you disturb significantly reduces the carbon footprint of your project. All of the carbon sequestered by the native landscape is released when the ground is disturbed. By selecting a previously developed lot, you won't be the first one to disturb centuries of carbon sequestration in the soils and native landscape on the site. Developing your site to protect the environment and biodiversity also means that you avoid future tax assessments to remediate environmental and ecological damages. Those activities not only cost money, they also carry their own carbon contributions.

Size Matters

Reducing the size of your home can have the most significant impact on its total carbon footprint. Larger homes take more materials to build and, depending on how well they are built, may take a lot more materials to maintain over time. Since each of these materials and the labor associated with installing them have their own carbon footprint, larger homes have higher embodied carbon. The size of your home also affects the energy and water used to operate it, and resources consumed to furnish it, keep it clean and maintain it over time. Therefore the size equally impacts the carbon related to all of those activities over the life of the home. So smaller homes, designed for efficiency with more flex space, accessibility features and able to accommodate changing needs over time, will have a significantly smaller carbon footprint over their lifespan.

Design

Carbon neutral design combines many of the principles covered in Chapter 3. A simple rectangular footprint, efficient floor plan and roof design, stacked stories, exterior dimensions on two-foot modules, providing design and framing details and using Advanced Framing techniques all result in resource efficiency, which means using less material during construction. The same is true for the design strategies that you use for the plumbing and heating and cooling systems installed in the home. Compact design provides efficient operations that can save energy and water over time. Wisely placing windows for passive solar design, shading, daylighting and natural ventilation strategies and placing all of the components of the HVAC system within the thermal envelope further reduces energy consumption. All of these efforts significantly reduce long-term carbon contributions.

Building Materials

Choosing building materials wisely, based on appropriateness for your project and long-term durability can significantly reduce your carbon footprint. One key step toward our carbon neutral goal is to build homes that use the best applied building science available to ensure the longest useable lifespan of every home constructed. Better construction practices that improve the durability of our home and lower maintenance requirements save resources over time. On your jobsite, using sound construction methods and quality-control practices, even those that might use more resources initially (like adding additional moisture, air or thermal layers), pay off in long-term carbon reductions from life cycle savings on energy, as well as reduced maintenance and repair cycles. All of these reduce a home's carbon footprint.

It is important to verify that sustainable harvesting and extraction processes were followed for every type of material used in the construction of your home. The extraction of everything from trees, aggregates, sand and stone to raw metals results in the release of sequestered carbon. These processes require large amounts of energy to perform, usually through the burning of fossil fuels, which produces greenhouse gases. Additionally, efforts to restore ravaged extraction sites also contribute to their carbon emissions. Companies with ISO 14001 certification have practices in place to reduce their carbon contributions by minimizing their impact on the natural environment and by how they manage their operations.

Transportation of building materials alone can be the biggest carbon-related component of the products used in the construction of your home. The absolute lowest carbon footprint would be represented by a natural building material that is sourced from the site itself. Beyond that, using products and materials harvested and/or manufactured locally, within 500 miles of your project, can do more than anything else to lower the total carbon footprint of your home's construction.

Look for products that have a higher percentage of locally sourced components. If you know the raw materials for the products are not available locally, look to manufacturers who have local production facilities. It takes much less energy to transport bulk raw materials to manufacturing facilities than it does to ship finished products one at a time to consumers, or even a truckload at a time to distributors. Many projects now focus their efforts on sourcing materials "Made in the USA."

Managing deliveries and scheduling trade contractors on the site to minimize unnecessary trips will also reduce the project's carbon footprint.[5] This also includes the number of miles driven by the project manager and the frequency of their site visits. If your contractor is also managing several other projects, they may need to leave several times each day; this can add a lot of unproductive transportation energy costs to each return trip. Of course, any power tools used onsite require an energy source, and the carbon contributions from that energy and the distance that it traveled to get to the site must also be calculated. There is much to be said for the old hand saw and hammer methods! Panelized, modular and engineered assemblies also have a lower carbon footprint than on-site assembly methods. ·

Also, the more renewable natural materials or salvaged, recycled or biobased products that are used on your project, the lower the carbon count associated with materials. Salvaged materials have a close to zero carbon footprint, since any carbon associated with reuse or repurposing an existing product would come from transportation. Recycled-content materials have the carbon contributed by the processes involved in recycling operations, but this is still considerably less than that released during virgin resource extraction and processing. Products made from rapidly renewable materials, such as bamboo or cork, represent less carbon than other alternatives sourced from old-growth trees. Products that require less finishing represent a reduced carbon contribution initially even over those with very durable finishes, and produce less over their life cycle from maintenance and upkeep.

Constructing your home using materials that are pest- and weather-resistant, or using integrated pest control measures, can mean significant reductions in the carbon attributed to repair and replacement cycles over the life of the home. Selecting materials that have classic styles, while considering your long-term needs, will also reduce the carbon embodied in furnishing your home. Designing for deconstruction, which allows the reuse of materials over time, negates the carbon value of those salvaged resources.

Construction Waste

Landfills contribute a tremendous volume of methane gas to our atmosphere,[6] contributing to climate change. Any strategies that can reduce construction waste save natural resources, including the energy and water consumed to manufacture the products that generated the waste. Investing in the oversight activities necessary to reduce waste—accurate material estimates, reducing waste factors, Advanced Framing techniques and construction waste management—not only saves you money, it reduces the carbon contributions associated with the manufacture, consumption and waste disposal of excesses. Additionally, waste that must be transported offsite for recycling or to end up in a landfill adds carbon contributions from the transportation involved. Finally, claiming materials back for reuse or repurposing saves all of the embodied energies related to the extraction of new virgin resources.

Equipment and Systems

Investing in good-quality equipment and systems, properly sized and properly installed to meet your needs efficiently, is the key to reducing the carbon emissions associated with the energy and water needs of these systems. Remember, air conditioning and heating is typically the largest user of energy in your home, so select systems that have proven long-term performance records, like the variable-speed inverter models that have recorded over 30 years of high-efficiency operations in places like Japan.[7] This means relatively little need to buy another system for a long time!

Combustion appliances contribute to carbon generation directly from their operations. However, high-efficiency models substantially reduce that impact as they require less energy, so gain savings associated with energy generation and transmission over their life cycle operations. Heat pumps exchange heat rather than burn fuels, so they save those related greenhouse gas emissions.

Water heating is second only to HVAC in energy use, so going for the highest efficiency here is also a priority. Remember the energy-water connection: saving water also saves the related carbon from the energy required to treat the water to potable standards. If your household consists of one or two adults who take short showers and don't generate tons of laundry, you might find a tankless natural gas unit to be efficient. However, if you have a few kids, who are or will someday become teenagers, you should consider solar thermal or a high-efficiency tanked model (gas or heat pump). By installing an on-demand recirculation system, you can make sure that you are not wasting too much water down the drain. You can further support that by specifying that all water fixtures and appliances be EPA WaterSense approved.

Mechanical ventilation for fresh air should be delivered by a system that provides simple and efficient operations. Stay away from any system that requires substantial amounts of energy or complicated technology, as these will probably not last the test of time and will only serve to increase your carbon footprint. The same is true for humidification and dehumidification systems.

Capturing rainwater reduces the carbon associated with water infrastructure. Regionally based water treatment and wastewater processing require tremendous amounts of energy. In fact, in many municipalities, the largest consumer of energy is the water and/or wastewater utility company, given how much it uses for treating potable water and pumping water and wastewater to treatment facilities. Potable rainwater systems require relatively little energy to treat the water, very little piping and only a small pump, as compared to the size of regional water utility systems. The same is true for graywater and septic systems, so the amount of carbon associated with these onsite facilities is miniscule when compared to their big brothers.

Installing high-efficiency appliances and light fixtures also reduces the carbon emissions related to energy generation and transmission. Electrical and lighting control systems, as well as home automation systems, might seem like unnecessary gadgets to those who make the effort to manually control their own energy usage, but for busy families these investments can make huge differences to their energy consumption over time. These devices can more than offset the initial carbon embodied in the creation and installation of these products.

As mentioned previously, burning fossil fuels to generate electricity is one of the largest contributors of greenhouse gases to our planet's atmosphere. The actual embodied energy factors for the energy used in your home will fluctuate depending on how and where the energy that you use is produced. If the fuel is combusted at an electrical power plant, to produce electricity, and then transmitted over long distances to our homes, huge transmission losses are incurred. If the fuel is pumped through distribution lines in its raw state, such as natural gas, and then combusted onsite, greenhouse gases are still released during combustion but are effectively less, as transmission losses are minimal. With hydroelectric power, significant greenhouse gas emissions result from the decay of trees and other vegetation flooded to create the reservoir needed to service the power plant. On the other hand, renewable fuel sources, like solar photovoltaic and wind, and non-combustion heating sources, like geothermal and solar thermal, do not generate any greenhouse gases, except those related to the small amount of electricity needed to run any electrical pumps that are required for their operations. In fact, any type of onsite power generation (e.g., burning wood, oil or propane) has a lower carbon footprint than grid-supplied power.

Health and Environment

In the United States, outdoor emissions related to certain volatile organic compounds (VOCs) are regulated by the Environmental Protection Agency (EPA), for the most part to prevent the formation of ground-level ozone, what we commonly refer to as smog. Although it is important to note that not all VOCs that impact indoor air quality are considered potential contributors to ozone, most of those that are released into the air occur during the manufacture or everyday use of products and materials, including chemical solvents used in building materials.[8] Just as off-gassing can occur over time to impact indoor air quality, the off-gassing of VOCs resulting in ozone can occur over several years from certain chemicals. These VOCs are included within the group of greenhouse gases that define carbon contaminants of our atmosphere. Using life cycle assessment (LCA) tools that provide testing and reporting of these constituents, we can reduce our carbon contributions from their use in the construction and operations of our projects.

Outdoor Living

Landscape design should incorporate native and adapted plant selections with good-quality soil to reduce water requirements. Replacing large turf areas with alternative ground covers, mulch or rock beds reduces the use of mechanical lawn equipment. These strategies reduce the carbon release from ongoing maintenance activities, as well as from the energy/water equation that we discussed earlier.

Of course, planting your landscape, especially adding native trees, can provide additional onsite carbon sequestration over time. Green roofs[9] and green walls can significantly increase this contribution. Trees also help to shade the structure, reducing cooling loads. Additionally, tree shading of hardscapes (patios, walkways and driveways) can help to reduce urban heat island effect, which can raise ambient temperatures and further exacerbate the need for mechanical cooling.

Designing site drainage that minimizes erosion of native soils and the need for supplemental irrigation will also protect the integrity of the site's sequestered carbon. However, much of our native soil has been depleted of nutrients. So, when restoring soils, soil amendments play an important role in carbon sequestration. Adding organic amendments improves the ability of vegetation to lock in nutrients in their root structure by taking in more carbon. Adding compost to your soil not only helps your plants, it helps the planet!

The Sustainable Sites Initiative (SITES) is a collaboration of the American Society of Landscape Architects, the Lady Bird Johnson Wildflower Center at the University of Texas at Austin and the United States Botanic Garden. After years of researching and analyzing and developing practices for sustainable site management, this organization has created voluntary guidelines and a 4-star rating system that encompass all aspects of sustainable land design, from site selection through landscape maintenance. The goal of the program is to encourage sites to be carbon neutral.

Finally, adding outdoor living space for cooking, showering and relaxing reduces heating, moisture and occupant loads from indoor mechanical systems during mild seasons when these spaces can be enjoyed. In turn, you reduce the related carbon contribution. Using your propane outdoor grill or solar oven can also contribute to reducing your carbon footprint.

Green Bling

Inefficiencies can also come in the form of green bling, not only wasting your money, but also contributing a carbon load that provides little, if any, benefit to your projects. For a truly affordable green home, you need to look at housing from a minimalist perspective, analyzing the necessity of each aspect or component and finding creative ways to use less. Keeping up with the Joneses not only wastes resources and adds unnecessary costs to your budget, it also contributes carbon for all those products that provide little benefit.

You need to analyze what your real needs are and how you can do your part by not purchasing things that you don't need or that don't fit your long-term goals. If you invest in passive strategies and more durable, high-performance materials and systems, you can make a significant reduction in the life cycle carbon associated with both your home's construction and its lifetime operations. At the same time, the strategies that we employ must also effectively lower the size of the onsite generation system (solar photovoltaic panels, wind turbines, etc.) that, in and of itself, has a carbon footprint. So, when we are able to meet our energy needs through a smaller system, we also get to our zero carbon operations goals quicker.

Keeping It Green

And, of course, "living green" reflects all of the daily activities made by our choices that contribute to our carbon footprint. Keeping your home green is about making choices to continue to reduce natural resource depletion and safeguard the environment. Many of these strategies can be used to reduce your consumption of products and thereby contribute to reducing your carbon contribution. Again, the goal is to reduce your carbon footprint not only for the initial construction of your home, but also over its life cycle of operations and maintenance. Sound maintenance, repair and replacement practices are necessary to keep your home operating as efficiently and durably as possible. Whether performing periodic self-inspections or having regularly scheduled commissioning and service work performed by qualified contractors, keeping systems—especially high-performance equipment—working at its intended capacity, is important.

Of course, that means following the same regimen of sourcing local materials for maintenance and repairs, and selecting durable, long-lasting materials. Making wise purchasing decisions on everything you buy based on total cost of ownership and reliable payback calculations, while choosing classic styles, will save you money over the long term and negate unnecessary carbon contributions from fad redecorating. Additionally, green power choices offer significant savings over the carbon contributions related to fossil fuel generation.

The choices that you make *after* you have moved into your home can have the greatest impact on your carbon footprint. There are a number of lifestyle carbon calculators available that can help you make better choices.[10] These calculators assign values based on automotive and airline travel, the types of foods that you eat and where they come from and how much waste and recycling you contribute.

We each must take the time to recognize opportunities to change our bad habits. Living in a community that offers amenities such as shopping, recreation or other community services within walking or bicycling distance doesn't mean anything if you still use your car to run all of these errands. Setting your thermostat on 72°F in the summer will not allow your system to operate efficiently. Running the dishwasher after every meal will not save energy or water. Taking 20-minute showers does not conserve water nor does installing a drought-tolerant landscape if you set your irrigation system to operate daily. Community recycling programs will not work if you continue to throw everything in the trash bin. All of these also increase your carbon footprint, so here are some additional efforts that you can make to conserve resources and lower your carbon footprint.

Use of our cars can be one of the largest components of our carbon footprint. In metropolitan areas, mass transit systems are available; many of these use alternative cleaner fuels like propane, electricity, biodiesel or natural gas. For those who live outside metro areas, you can ride the commuter train to work and support car share programs while in town. Some people ride their bicycles to work and then carpool to meetings with others when possible. There is also a trend toward living in multi-use developments that encourage pedestrian and bicycle traffic for extracurricular activities, like shopping and recreation. All of these choices reduce the carbon emitted from automotive exhausts.

Support local businesses that manufacture products, since transporting products from outside sources to local markets uses fossil fuels, contributes to pollution and climate change and adds cost to those products, as well as sending your money out of your local area. Supporting farmers' markets and local CSAs (Community Supported Agriculture) that provide locally grown food is a good start. Locally sourced food contributes to better health and is better for the environment, since it means that we are not purchasing chemically processed food and shipping it from around the world with greater spoilage and waste.

Practice the same vigilance in shopping and waste reduction for your daily needs as you did for your construction or remodeling project. Think about long-term use of what you buy and whether you could spend a little more to buy durable, long-lasting products. Don't buy what you don't need, so avoid compulsive spending both while picking up your listed items in the store as well as at the checkout counter. Keeping a running list of items that you need and errands that you must run and buying in bulk are two ways that you can reduce the number of trips you make. Preplan, so that when you have to make that trip to the market, you'll also have time to stop and take care of any errands within close proximity to your route. This saves gas and wear and tear on your car, as well as reducing automotive-related pollution and overall commute time in your week.

Choose to buy products that are reusable, like cloth cleaning towels, rather than disposable, like paper products. The best way to make this transition is to keep a mindful eye on what ends up in your trash receptacle. Investing in a faucet-mounted water filter and reusable water bottles for each member of your family makes more sense than continuing to buy single-serving disposable water bottles. Investigate your local resources for recycling. If you are not fortunate enough to have community supported curbside pickup of recyclables, purchase some dedicated bins for storing these and add this to your errand route periodically. Of course, if you can repurpose items yourself, like using old milk cartons for starting seedlings for your garden, even better.

Of course, energy and water conserving efforts are an integral part of keeping it green, which means replacing burntout light bulbs with high-efficacy CFLs or LEDs. You can practice personal care in how you use water, making sure to turn off the tap while you're brushing your teeth and scrubbing the bugs off while washing your car. Always remember the energy-water connection: saving water provides double the reward plus reduces the carbon from that production and distribution!

All of the energy-saving recommendations in this book can contribute to reducing peak loads, but much more can be done by timing as many of your activities as possible to occur during non-peak periods. The time of day and/or time of year that you consume power can significantly affect the related greenhouse gas emissions. Peak demand periods, especially in hot climates, can maximize grid-supplied power from dirty sources, and eventually this can increase the need to build more fossil fuel-burning power plants to meet demand.

The problem with most renewable energy is that we currently have no mass storage capabilities, which means that we still have to keep the coal and nuclear plants turned on in case we need them. These large power plants just can't be turned on and off with the flip of a switch. So they continue to produce power and, therefore, continually produce greenhouse gases.

The only way to break this cycle is to rebalance demand to meet more of it during non-peak periods. Programming your dishwasher to run in the middle of the night while you are asleep, or waiting to do chores or housework until the morning or after dinner can prevent related peak demand issues. It can also save that extra carbon contribution, and, if your utility has a demand charge, you can also save some big dollars on your utility bill.

Many design strategies can contribute significantly to the amount of energy required to operate your home, so if you have the option of using passive systems in your home, use them. Even if you do not have these features, you can still make choices in your behavior or changes to your home that will help to lower your utility costs. These could be as simple as opening the blinds to let in natural light to provide general illumination and then limiting the use of electrical lighting to specific tasks that you need to perform. Install a clothesline and ditch the clothes dryer except in periods of extended rain when you just can't wait for sunshine to do the laundry. Open the windows and let the natural breeze cool your home in the milder seasons or turn on the ceiling fan and turn up the air conditioner a couple of degrees. Add awnings over windows to provide exterior shading or plant trees to shade entire walls. If you are remodeling, replace refrigerators, dishwashers and clothes washers over ten years old with ENERGY STAR- and WaterSense-rated models. Replace older faucets, toilets and showerheads with low-flow rated models.

Plan your family's schedule so that you can all sit down to dinner together. Not only can this provide quality time together, it also means you save on energy and water by cooking for a larger group at once. In fact, consider expanding to larger family gatherings or inviting friends and trading off locations, so everyone saves more. Opt for using more energy-efficient cooking methods, like slow cookers or even solar ovens. Avoid microwaving, as making anything happen at mega-speed also means using mega-energy. Only run the dishwasher or clothes washer when you have a full load. And try washing your clothes in cold water. Use the energy-saver setting on the dishwasher, so the dishes dry without the added heat feature. Use smart outlet strips to control phantom loads and add

an insulation blanket to your tanked water heater. Use energy management software to monitor your home's energy use so that you
can continue to reduce your usage. All of these activities reduce the
carbon footprint of your home's operations.

Better to adjust to these small changes now than have to make
harder ones later when energy and water utility costs are unafford-
able or unavailable because the resources have been depleted. As
our population on this planet continues to increase, we can choose
to use what we have more conservatively, endure rationing and
blackouts or resort to building more fossil fuel-burning power
plants, contributing more carbon, leading to more extreme weather,
creating a vicious cycle that we can't get away from. We choose *conservation.*

All the Other Zeros Contribute to Zero Carbon

Efforts to avoid peak demand spikes can include using renewable
energy sources to supplement high demand, further lowering our
carbon output. Achieving net zero energy through the use of onsite renewable power generation does not completely eliminate the
energy piece of our carbon footprint pie, though, since the grid is
still used for transmission and storage over the life of the home.
To get completely to zero carbon, the home would need to be off-
grid entirely, for energy (Chapter 11), water (Chapter 12) and waste
(Chapter 13).

Our Carbon Reduction Plan

To achieve the lowest carbon footprint from construction, operations and maintenance of your home, these are the strategies that
have been discussed in depth throughout this book:
- Select a previously developed infill site (could be a remodel on
  existing slab) to reduce carbon emissions by using existing community infrastructure versus new sprawl development (streets,
  utilities and community services).
- Locate in an existing community for shared amenities (shared
  parking, walkable distances, community center, community

garden, pool, etc.), or a mixed-use community (live/work/play shopping, community resources and mass transit) for additional reduced carbon emissions from auto use.

- Reduce home size in order to lower emissions, using less material to construct a smaller footprint, including furnishings.
- Design custom passive solar, passive ventilation and floor plan layout to take advantage of passive heating and cooling strategies.
- Design the home with a central core plumbing design, central HVAC system design and managed electrical system design.
- Eliminate the garage by including a car port.
- Use sound building science and building-as-a-system integrated high-performance systems to achieve net zero energy (or at least capable).
- Reduce the quantities of materials used through design efficiencies, Advanced Framing, panelized construction and construction waste reductions.
- Limit the area of the site that is disturbed, fencing off areas to protect the natural landscape and staging onsite work crews within the boundaries of the detached garage and outdoor living areas.
- Reduce embodied energy by sourcing from ISO 14001-certified manufacturers, using salvaged and local materials (within 500 miles) for construction and using materials that serve more than one purpose.
- Stockpile and reuse native soil, restore habitat and landscape with native plants.
- Reduce the burden on water treatment and pumping infrastructure through indoor and outdoor water conservation measures and graywater/rainwater collection systems.
- Add outdoor living space and amenities, including a clothesline, outdoor shower, summer kitchen and screened porches or sunrooms (where climate appropriate) to reduce the mechanical systems required to maintain comfort inside your home.
- Reduce normal maintenance and repair cycles, as well as from

upkeep and cleaning, by using durable and long-life, low-maintenance materials and quality construction methods.

- Reduce landscape maintenance by restoring native landscapes, reducing turf area, adding soil amendments and mulch and limiting mowing and other maintenance activities.
- Use natural pesticides, fertilizers and cleaning products with less off-gassing than traditional chemical products.
- Perform continued commissioning to maintain optimum performance of the structure and systems.
- Commit to making behavioral changes to avoid peak demand and ozone warning usage periods.

For those of you who are intent on getting to zero carbon in your construction project and the operations of your home, there are resources out there that can assist you. The Athena Sustainable Materials Institute (calculatelca.com) has developed a couple of extremely useful tools for architects and homebuilders that measure impacts from various building materials and methods. Their Eco-Calculator for Residential Assemblies and Impact Estimator allow professionals to analyze building assemblies and entire buildings based on life cycle assessment methodology. The Build Carbon Neutral calculator (buildcarbonneutral.org) will estimate the carbon released and embodied energy in construction. The Green Footstep calculator, from the Rocky Mountain Institute, (greenfootstep.org) is designed to allow architects and builders to model carbon from both construction and life cycle operations based on locale, building type and size, energy efficiency and age of construction. These entities offer support services for the use of their tools.

CHAPTER 15

Zero Cost Premium

This book was written to help you understand the total cost of homeownership and how to use specific strategies to achieve a high-performance green home for no additional life cycle cost. Throughout the book, you have seen the symbols for "No Cost Green" and money-saving green strategies. It is difficult to put actual dollar values on each of those recommendations, as costs can vary depending on market conditions and local availability of materials and skilled trades. However, these strategies, if incorporated into your project at the appropriate time, can both save you money initially and give you an affordable green home that continues to provide a hedge against unforeseeable and rising operational and maintenance costs over time.

In this chapter, we will build a model comparing the total cost of a traditionally built house to a home scenario that we have developed based on many of the recommendations we have made in this book in order to represent a zero cost premium scenario. Our goal is to build a truly green home for no more cost than what our total amortized cost of ownership would be in the traditional home over a fifteen-year period. This model incorporates the same types of strategies used in our net zero energy model (Chapter 11), plus additional features that were used to balance our cost savings and provide our desired return on investment of fifteen years or less.

We'll start by recapping the first section of the book again, this time with an eye to how each decision made will impact your budget.

Reducing Your Total Cost of Ownership

In Chapter 1: Location, Location, Location, we recognized the external costs of housing related to site selection. We recognized that location and site features contribute costs to our household budgets, including the long-term costs for:

- Commuting expenses and value of commute time lost
- Property taxes associated with sprawl developments, expanding infrastructure and community services
- Loss of and damages to ecosystems, air quality, farmland and natural habitats due to sprawl development and related pollution
- Utility, maintenance and repair costs when selected sites do not provide passive benefits (passive solar heating, passive ventilation or natural drainage) or are too risky to use for long-term housing

It is impossible to assign a cost to air pollution, environmental damages and the impact on wildlife or our cost of food. But it is important to remember that we, as a society, share those costs. Remember, we externalize many costs in other forms, and the full extent of their value may not be determined for decades into our future, so they cannot be reflected in our modeling here. Some of those costs are being subsidized, to some extent, through state and federal tax incentives. Others are managed through the efforts of charitable non-profit institutions, while others remain issues that some future generations will no doubt ultimately have to deal with. Nevertheless, we should recognize that these costs are, to some extent, the result of the choices that we make in terms of housing location, contributing to land development and commuter-related pollution and impacting our base tax rates for local, state and federal tax structures. For budget comparison purposes, our baseline will be built in a bedroom community suburban setting and our zero cost premium home on the recommended urban infill lot.

Reducing Your Construction Costs

The next four chapters of the book focused on strategies that can lower our base construction costs. In Chapter 2: Size Matters, we discussed the fallacy of comparing home prices per square foot, since all home budgets include common elements (kitchens, baths, HVAC, water heating) regardless of the home's size. But we did recognize the cost savings from building a smaller home with flexible use of space features, both in the construction budget as well as in furnishing, cleaning and maintenance of a home over time. Smaller homes typically take less energy to operate, so those utility cost savings can really add up. In our cost model, we will use the same two-story home with 1,862 square feet of conditioned space as the net zero energy home model discussed in Chapter 11.

In Chapter 3: Design, Design, Design, we recognized how much money we saved by incorporating resource-efficient design strategies. Again, we will have the same rectangular footprint of 28 by 40 feet as our net zero energy home, which reduced the shape factor to 16.5. In our model, the traditionally built home has a chopped-up roof design for "architectural interest," which would make it very complicated and expensive to build, and also add considerable costs to maintenance and repairs over time due to its increased susceptibility to leaks. The roof of our zero cost premium home is redesigned for simplicity to improve durability (minimizing potentially leaks in the future) and to maximize our ability to collect rainwater and provide a straight south-facing roof plane, allowing us to plan for solar access. Both our traditional home and zero cost premium home will have slab-on-grade foundations engineered for the site.

Of course, the design of our zero cost premium home includes passive solar, passive ventilation, daylighting and correctly sized overhangs, both for shading and protecting the walls, windows and doors for improved durability in our home design. As you recall, the baseline home in our energy model had an 18 percent window-to-conditioned-floor-area (CFA) ratio, which we were able to reduce to 8 percent in our net zero energy home. We use this same strategy in our zero cost premium home. We have also designed on two-foot framing modules and stacked its stories using Advanced Framing techniques and engineered trusses, including energy trusses at the roof. We are using the same net zero energy strategy of a sealed unvented attic, bringing the mechanical systems within the thermal envelope.

We improved the room layout to focus around a central core, stacked plumbing design, with graywater reuse and a central HVAC system location. The kitchen location on the north wall meant that we could exhaust heat with the prevailing southerly breeze in the summer, yet retain it during the winter. All of these design features can significantly lower our home construction, maintenance and operating costs.

We followed through with a lighting design that would allow us to utilize task lighting to supplement daylight, when it was available. We then looked at different general illumination alternatives to fit the needs of the space. We carefully planned our structured wiring to provide the ability to use smart panel technology to reduce vampire loads from non-essential sources when they were not in use and to work with other smart controls that we planned to include in the finish out.

These design efficiencies significantly contributed to lower base bids on the construction of the home. Savings from reduced framing materials and labor, central core plumbing and right-sized HVAC system installation and durable finish materials were used to pay for other improvements that will pay for themselves over time. This more than offset the cost of paying a qualified local green home designer for design services and an energy modeling expert for guidance on choosing the right alternatives to meet our net zero energy performance goals. Our integrated project team provided synergy to the design process, assuring us that we would achieve overall whole-building performance and long-term durability, as well as optimizing the size, design and performance of each individual system. The result was a design that will provide everything that we *need* in a home, while leaving us money on the table to put toward the upgrade in systems and finishes that we *want*.

Our traditional home is built to code specifications, as described in Chapter 11. Following the strategies discussed in Chapter 4: Building Materials and Chapter 5: Construction Waste, our project team gave us expert guidance on the best materials, systems and finishes to achieve our remaining zero cost premium home goals of durability, low maintenance, energy and water efficiency and healthy indoor air quality. This included 100 percent durable flooring, green cabinetry, ENERGY STAR- and WaterSense-rated fixtures and appliances and low-VOC adhesives and finishes.

Details of the building specifications for our zero cost premium home model are as follows:

- One-inch rigid polyiso foam board insulation at stem wall of foundation (exposed area and six inches below grade), R-5
- Two-by-six exterior (engineered) wood-framed walls using Advanced Framing techniques, including 24 inches on-center stacked floors
- Engineered floor and roof trusses clear span all non-load bearing interior walls
- Blocking for grab bars at all baths
- Two-by-six interior plumbing walls with non-load bearing two-by-four interior walls, 24 inches on-center
- ENERGY STAR Grade I insulation: open-cell spray foam insulation, R-15 at walls
- Spray foam insulation at roof rafter, R-21
- Casement windows, < 0.4 U-value, < 0.3 SHGC (climate specific per EWC)
- Half-inch structural insulated sheathing (R-3), taped at all joints and openings
- Insulated headers over window (two-by-six with minimum of two inches rigid foam insulation)
- Woven house wrap (second layer drainage plane), flex wrap flashings (shingle layered) around all window and door openings
- Fiber-cement siding and exterior trim
- Twenty-six-gauge standing seam or concealed fastener metal roof, galvalume, SRI > 29

Healthy and High Performance

In the following chapters, we introduced materials and methods to improve indoor air quality and energy efficiency in our homes. Following the recommendations in Chapter 6: Equipment and Systems, we chose to install a simple fresh air system designed to meet ASHRAE 62.2 as per the national codes. The model that we selected is an Energy Recovery Ventilator (ERV) that preconditions the incoming fresh air. Our ductless mini-split HVAC units also provide humidity control. In selecting mechanical systems, we elected not to include any combustion appliances in our zero cost premium home, completely eliminating any concerns for carbon monoxide and combustion backdraft safety. We also specified that all of our bath exhausts would be either humidity sensing or on timers. We specified non-toxic termite and pest products and an approved metal mesh to be applied to all foundation penetrations. We are also sealing all wall, floor and joint penetrations to prevent pest access. We are not in a risk area for radon soil gas contamination, so we have not addressed this.

In Chapter 7: Health, we selected the building materials for both inside the walls and interior finishes with the following specifications:

- Cabinetry made with non-urea-formaldehyde binders and finishes
- Linoleum flooring in baths; entire first floor stained concrete; wood flooring in upstairs
- Solid-surface countertops
- Engineered (finger-jointed) interior trim (base and door casings, urea-formaldehyde free)

Of course we are specifying all low-VOC products for both building materials and finishes in our zero cost premium home. In the box on the following page we have included the specifications for all products in those categories.

If you recall, in Chapter 11, the design of our net zero home eliminated the attached two-car garage that is found in the typical

Paints

Note: all paints must contain no prohibited ingredients and meet Green Seal GS-11 Standard as follows:

Product Type	VOC Limit (grams/liter)
Flat Topcoat	50
Flat Topcoat with colorant added at the point-of-sale	100
Non-Flat Topcoat	100
Non-Flat Topcoat with colorant added at the point-of-sale	150

All caulk and sealants must comply with Regulation 8, Rule 51 of the Bay Area Air Quality Management district (= or < 250 grams/liter)

All adhesives must comply with Rule 116B of the South Coast Air Quality Management District as follows:

Application	VOC Limit (grams/liter)
Wood flooring adhesive	100
Subfloor adhesive	50
Ceramic tile adhesive	65
Drywall and panel adhesive	50
Multipurpose construction adhesive	70

All stains and finishes intended and labeled for use on wood and select metal substrates meet Green Seal GS-47 Standards as follows:

Product Type	VOC Limit (grams/liter)
Stains	250
Sealer	200
Waterproof sealer	250
Low solids coating	120
Varnish	350
Lacquer and pigmented shellacs	550

home, and instead elected to build a separate two-car open-air carport. We do this in our zero cost premium home to eliminate any risk of pollutants entering our home from automotive exhausts or pesticides, fertilizers or various chemical compounds typically stored in the garage area, but it is also much less expensive to build a carport than an enclosed garage. Also, since we live in a predominantly hot climate, but do occasionally enjoy a few cold winter days, we decided also to locate our only fireplace in that outdoor living space, improving the overall thermal performance of our home, as well as reducing possible contaminates to our indoor air quality. This also represents a considerable cost savings since it does not require all the special flashings and materials needed for an indoor fireplace.

In Chapter 9: Green Bling, we discussed using some of our cost savings from our base budget cutting strategies to fund upgrades. We have used our savings to upgrade to the following systems and finishes in our zero cost premium home:

- Four ductless mini-split HVAC units (12,000 BTU/hour downstairs, one 6,000 BTU/hour in each upstairs bedroom, and a 9,000 BTU/hour in the upstairs open living space), 19 SEER/ 12 EER
- ERV: minimum 0.35 air changes/hour
- Full house gutters piped to a 5,000-gallon rainwater cistern
- Solar-thermal water heating
- ENERGY STAR-rated light fixtures and ceiling fans
- WaterSense and CEE Tier 3 appliances
- Showerheads and faucets < 2 gallons per minute rated
- Occupancy sensors on lighting
- Six kilowatts of solar photovoltaic with energy monitoring

We also embraced the benefits of adding outdoor space, as discussed in Chapter 8: Outdoor Living. This would help to reduce our internal heating and cooling loads and moisture loads, as well as give us the health benefits of communing with nature and supporting wildlife. With this extended outdoor living area, we end up with 2,251 square feet under roof (not including our 400 square foot carport), making up for our reduction of conditioned space to 1,862 square feet. The added area under roof includes a nice 18 × 10 front porch as a passive feature on the south side. This and the remaining space gives us plenty of room to add a summer kitchen and dining area, and an outdoor sitting area for entertaining.

Some of our other outdoor features that lower our total cost are:

- Native topsoil stockpiled and reused
- Twenty-five percent compost soil amendment for top six inches of all grade
- Grade to direct all non-captured stormwater to rain gardens and planting beds using berms and swales
- Planting beds with two inches of organic mulch and all plants from drought tolerant native/non-invasive plant list

These efforts will minimize our landscape watering needs. Since our site is located within a municipality that prohibits the potable use of rainwater if you have access to the city's treated water system, our rainwater collection system will be used only for our outdoor water needs. Hopefully at some point in the future, all cities will recognize the benefits of individual capture and treatment systems, and let us expand our storage capacity and install the necessary equipment to provide all of our potable water as well.

Putting It All Together

There is no limit to the number of scenarios that could be created using different building methods, materials and systems. Many builders rely on years of experience with what has worked or not worked in relation to structural and water infiltration issues, durability and, of course, homeowner comfort. Some experiment with different combinations, continually trying to improve overall home performance or, at least, limit their liability for callbacks and warranty claims (and customer dissatisfaction). Building products themselves are continually changing, with both new products and product improvements being field tested every day. Each home's design and individual combination of selections will create unique variables to affect both its cost and its return on investment.

And, of course, each home can also perform differently due to site conditions and occupant behavior. It is, therefore, impossible for this book to provide you with exact recommendations on what to do, what products to use or what your cost differential will be. Those who promote green building do offer case studies that reflect what they were able to achieve with their projects using a particular combination of strategies. So, before we delve into the spreadsheet for comparing the total cost of ownership of a traditionall built home and zero cost premium home, let's take a look at a cas study done by a third-party Fortune 100 firm comparing a high performance home to a home built to common building codes.

Case Study

BASF manufactures a broad range of building materials and ha dedicated resources to promote high-performance building prod ucts and improved building science applications through thei Center for Building Excellence. Having partnered with Environ ments for Living (EFL) in developing the BEYOND Applied Build ing Science certification specifications (see Figure 15.2 for the Plar Analysis Summary), they were able to reduce their Home Energy Rating System (HERS) score from 98 to 50! They did this at a cos premium of only $3,378, with a modeled recovery of that invest ment over a 3-year energy operational cost savings.

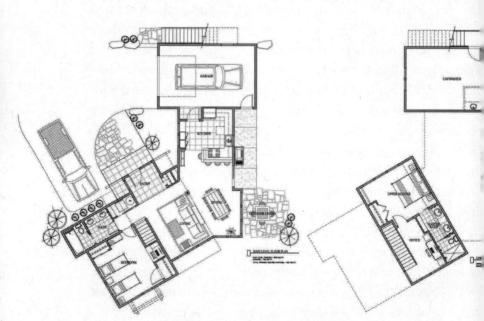

FIGURE 15.1. BASF Case Study Floor Plan. Credit: BASF.

	BASE IECC 2006	BASF BEYOND Program
THERMAL ENVELOPE		
Slab	Uninsulated	match existing
Exterior Wall	R-13	**R-18= R-13 + R-5 C.I.**
Advanced Framing	No	**Yes**
Window U-Value	0.33	match existing
Window SHGC	0.30	match existing
Ceiling	R-30 Unconditioned Attic	**R-21 oc SPF**
Radiant Barrier	No	No
House Leakage Target	Visually Inspected	**2 ACH 50**
MECHANICAL		
Heating System	80 AFUE	**94 AFUE**
Cooling System	13 SEER	**16 SEER**
AC Size	5 tons	**2.5 tons**
Equipment Location	Unconditioned Attic	**Conditioned Attic**
Programmable Thermostats	Verify	**Yes**
Supply / Return Duct R-Value	R-8	**NA**
Supply / Return Duct Location	Unconditioned Attic	**Conditioned Attic**
Duct Testing	Verify	**Exempt per IECC 2012**
ELECTRICAL		
CFL % or LED %	Verify	**95%**
Installed Appliances	No	**ENERGY STAR**
Misc. Electric Loads (MELs)	No	**Yes**
PLUMBING		
Water Heater EF & Type	0.62 EF Gas	**0.92 EF On-demand Gas**
Plumbing Distribution	From Garage	**Attic-sealed Combustion**
Hot Water Pipe Insulation	Verify per code	**Yes, R-4**
Low-Flow Plumbing Fixtures	No	**Yes**
COST ANALYSIS		
Estimated Upgrade Cost		**$3,378**
Annual Utility Costs	$1,700	**$591 3yr ROI**
HERS Index	98	**50**

FIGURE 15.2. BASF Case Study. Specs. Credit: BASF.

To provide more details into the cost trade-offs for achieving this low incremental investment and fast payback, the following spreadsheet indicates the changes made to the base 2006 IECC building specifications.

Items shown as credits indicate the deletion of the originally specified 2006 IECC code-compliant home specifications. Charges immediately following each of the credit line entries indicates the BEYOND building specification that replaced the credited item. As you can see, some of the major changes include:

BEYOND. Estimated Upgrade Costs				
	Current Specification Upgrade Specification	Unit price	Area or Quantity	Upgrade Cost
THERMAL ENVELOPE				
Above Grade Walls Continuous (not on garage wall)	Housewrap	$0.20	1632	(326.40)
	R-5 Continuous Faced	$0.53	1632	864.96
Continuous Air Barrier	Joints and penetrations	$0.20	163	32.60
Advanced framing	2x4 @ 16"oc	$0.41	1921	(787.61)
	2x4 @ 24"oc	$0.27	1921	518.67
Roof/ Ceiling	R-30 blown in	$0.32	2268	($725.76)
(includes Ignition barrier)	R-21 oc SPF pitch	$1.55	2542	$3,940.10
Attic hatch into conditioned attic	Insulated Attic Hatch	$70.00	1	($70.00)
LF Ridge	Remove Ridge Vent	$3.70	72	($266.40)
Soffit Vent	No vented soffits	$60.00	1	($60.00)
Gable End Walls	Add Gable Wall Insulation	$1.55	480	$744.00
MECHANICAL				
Ducts in conditioned space	No duct testing required	$200.00	1	($200.00)
Equipment Size	5 tons	$8,000.00	1	($8,000.00)
	2.5 tons	$5,750.00	1	$5,750.00
Cooling Efficiency	13 SEER			
	16 SEER	$1,000.00	1	$1,000.00
Programmable Thermostats	Verify Existing			
	Yes	$75.00	1	$75.00
Mechanical Ventilation	No			
	Yes	$150.00	1	$150.00
Short Duct Runs	See Manual J reports	$1.00	100	($100.00)
Reduced Duct Insulation	R-6 in conditioned attic	$150.00	1	($150.00)
ELECTRICAL				
CFL % or LED %	Verify Existing			
	95%	$50.00	1	$50.00
PLUMBING				
Water Heater	0.62 EF Gas	$500.00	1	($500.00)
	0.92 EF On-demand Gas	$1,200.00	1	$1,200.00
Plumbing distribution	No			
	Yes- centralized in attic			
Hot water pipe insulation	Verify with code			
	R-4	$1.91	125	$238.75
Low Flow Fixtures	No			
	Yes			

BEYOND. Estimated Upgrade Cost $3,378

- R-5 continuous insulated sheathing, spray foam insulation at rafters in lieu of blown in the attic floor
- Sealed attic assembly bringing HVAC equipment and ductwork within the thermal envelope. Higher-efficiency HVAC water heating system and pipe insulation. Improved HVAC and plumbing design and distribution
- Fresh air ventilation

It is important to note that, due to the improvement in the thermal envelope and HVAC design, the size of the HVAC system was able to be reduced from 5 tons to only 2.5 tons, a savings of $2,250. Along with the other envelope improvements, the combined savings are almost enough to pay for the spray foam upgrade and resulted in reducing the annual operating energy requirements by the home by almost 50 percent.

Now, back to the strategies that will get us to the Zero Cost Premium Home.

The "Convertible" House

Our final strategy for our zero cost premium home model came from our thoughtful consideration of how we might use this home over time. Having carefully selected a location that offered us the benefits of being able to age in place, one with all the support services that we would need as we get older within a short commute, we recognized that our need for housing would change over time. When we are young newlyweds, planning our future, we want a home large enough to raise a family in. We think the three-bedrooms–two-bath model of the traditional home fits that scenario. However, what will we do with all of that space—in fact, what use will we have for the entire second floor—after the kids are grown and gone?

We could sell the home, but by that time, say 25 years from now, what would our options be for staying affordably in the same neighborhood? By then we could have this home paid for and not be willing to go into debt again in our later years. So, what if we designed

FIGURE 15.4. The convertible house.
Credit: Studio Momentum, Travis Young Architect.

our home to be able to convert it into stacked condominiums at that point in our lives?

Some analysts believe that as our populations continue to grow and we need to increase the density of our urban environment to accommodate more people, we will see this type of remodeling occurring more frequently. In fact, this supports the belief that our homes will continue to get smaller, as is already seen in some of the more densely populated cities. There may be a day when the average American home is a two-bedroom condo in a dense urban location.

Also, with the increasing number of baby boomers who are downsizing,[1] the demand for smaller units to accommodate their needs is increasing. So we've decided to design and build our 1,862

Added wall to close off stairs

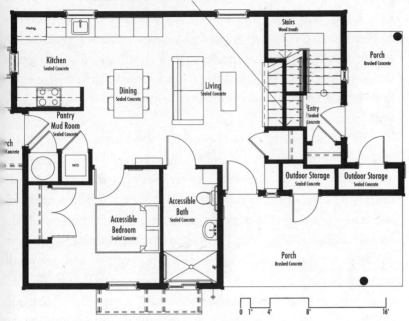

FIGURE 15.5. Convertible House: first-floor stacked duplex.
Credit: Studio Momentum, Travis Young Architect.

Install cabinets, fixtures, and appliances

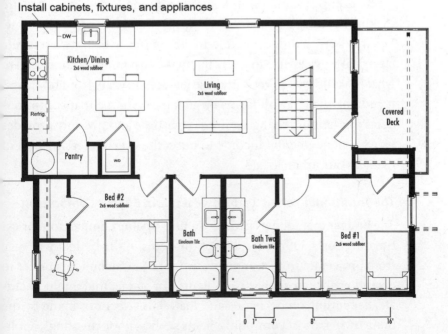

FIGURE 15.6. Convertible House: second-floor stacked duplex.
Credit: Image courtesy of Studio Momentum, Travis Young Architect.

square foot two-story home in such a way that it is pre-plumbed to be separated into two separate stacked condominium units at a later point in time. This means that, with the addition of a single short wall to close off the downstairs access to the staircase, the first-floor side patio now becomes the entrance to the upper condo unit.

The upstairs game room is pre-plumbed for a kitchen on the sidewall opposite the stairs, and as you may recall from Chapter 11 we already have a second water heater to service the upstairs. Both units will be wired to separate electrical boxes. The downstairs unit becomes the aging-in-place one-bedroom condominium while the upstairs becomes a separate two-bedroom condominium. After our children are grown and gone and we are middle-aged, we could move our own parents, who are now aging, into that first-floor accessible unit. After they are gone, we would occupy that space ourselves in our golden years, making available the upper unit for our kids to raise their own family or as a rental unit to supplement our retirement income.

By doing this planning before our home is ever built, we are making the most efficient use of our space and resources over time. Yes, it will cost a little more now to pre-plumb, wire for two separate electrical services and so forth, but that is insignificant compared to what it would cost to retrofit these features 25 years from now. And it certainly will pay off for us by giving us a home that accommodates our family's changing needs over time and possibly provides us with supplemental income to offset the ever-increasing cost of living in our urban locale.

The Traditional Versus Zero Cost Premium Budget Comparison

The budget that follows provides a line-by-line comparison for estimating costs. Of course, any of the items in the budget will vary from project to project. In many instances, homebuilders prefer to give allowances for the various items in a budget that might fluctuate, depending on decisions and materials selections made by the homeowner. Most commonly, items such as floor coverings, appliances, light fixtures, plumbing fixtures, cabinets and countertops and other interior finish materials are provided an allowance, and the homeowner is expected to pay for cost overages if they select specialty products or more expensive materials.

But other basic construction costs will also vary. A homeowner might see a particular brand and style of window that they prefer and a cost variance might result. Or site selection may cause variances in the cost of the foundation (on sites with heavier topography, slab vs. basement) or for sitework (e.g., cut and fill for driveways, foundation pads or clearing trees). The zero cost premium budget shown in the comparison is based on a project built in 2013 with only a few minor changes to its actual specifications. For simplicity, we decided to round off all of the cost entries to the closest hundred dollars, as our intent is just to exemplify our comparison of total cost of ownership over time, not to actually give you another case study. Remember, the goal of our zero cost premium budget is to deliver a truly green home that has no greater cost than a traditionally built home over a 15-year amortization schedule.

As you look at the comparison, you will notice that some costs of greening the traditional home did not change much, while others changed dramatically. For example, the poor, inefficient design of the traditional home caused framing costs to be very high for this size of home. This also affected the cost of the roof, due to the chopped-up roof design that resulted from a high foundation shape factor coupled with the off-the-shelf house plan's attempt to use the roof as an architectural feature. By tweaking both the design and the structure with Advanced Framing strategies, we are able to pay a skilled craftsman a higher labor rate, pay more for better engineered structural members, add insulated sheathing and still save money. On the other hand, though we budgeted much more for healthy interior finish materials, we have no way to quantify the return on our investment in terms of health benefits for that extra money spent.

Note that the costs of each home are presented as a sample estimate, not an actual budget for the projects. Pricing may fluctuate depending on contractor selection, specific materials and methods used and changing market conditions.

Standard (Baseline) Home		Zero Cost Premium Home	
Description	Cost	Description	Cost
1,862-square-foot (conditioned space) traditional-style, two-story with an attached 400-square-foot garage		2,251 square feet under roof: includes 1,862 square feet conditioned space plus 365 square feet covered for front and back porches with separate 400-square-foot carport	
Home plan: stock plan, not designed for any efficiencies	$1,800	Custom passive design, integrated team input, two-foot modules, with shape factor <24, energy modeling, including engineering fees	$ 9,800
Foundation: slab on grade, over-engineered due to no soils test, no site topography grade change	$22,500	Engineered slab on grade foundation and carport (same topography as baseline)	$17,900
Framing and lumber purchase: inefficient design, tradition stick-built wall assembly with garage	$53,600	Advanced Framing (turnkey), simplified design on two-foot modules, engineered floor and roof trusses, waste reuse, insulated sheathing, two-car open carport	$44,000
Windows: 18 percent glazing/CFA (standard dual pane) and exterior doors	$8,500	Windows: 8 percent glazing/CFA and exterior doors (ENERGY STAR LoE, U-value 0.4 or less, SHGC 0.3 or less including daylighting and ventilation applications	$5,500
Plumbing: not duplex, traditional design, standard efficiency gas water heater, standard fixtures	$14,800	Plumbing: central core design with separate connections first and second floors, solar thermal water heater, graywater reuse, WaterSense fixtures, less rebate, tax credit	$20,200
HVAC: inefficient design and layout, based on traditional minimum efficiency, located remote, 5-ton standard efficiency = 400 square feet/ton	$10,900	HVAC: four ductless mini-split units, ERV	$17,000
Electrical: standard wiring	$5,700	Electrical: structured wiring with separate service to upstairs and downstairs units	$8,000

Standard (Baseline) Home		Zero Cost Premium Home	
Description	Cost	Description	Cost
Roof: complicated design, 30-year shingles including garage)	$7,100	Roof: simplified design, 26-gauge standing seam metal, including carport, awning shading devices	$10,300
Insulation: R-13 batt wall and R-30 blown ceiling	$4,800	Insulation: spray foam R-15 at walls, R-21 at rafters, foundation insulation	$6,700
No testing performed: systems operational efficiencies unknown		Commissioning and performance testing	$700
Standard builder finish allowance: sheetrock/texture, cabinets, interior doors/trim, laminate countertops, carpet and tile flooring, mirrors, standard appliances and light fixtures, underpinning	$30,200	Finishes: stained concrete first floor, wood second floor, linoleum in baths, sheetrock/texture, green cabinets (first floor kitchen only), interior doors/trim, solid-surface counters, ENERGY STAR and WaterSense appliances, ENERGY STAR lighting (utility design), underpinning, termite treatment, gutters	$39,900
Landscaping and irrigation: purchased poor soil, traditional plant selections, spray irrigation system	$10,500	Landscaping: native soil, amended, and native plant selections, raingarden/erosion control, high-efficiency irrigation system	$14,000
Construction project management: trash service, porta-potty, electric and water service, cleaning, inspections, builder's risk insurance	$27,300	Construction project management: waste diversion service, porta-potty, electric and water service, cleaning, inspections, builder's risk insurance	$15,800
Total Base Construction Budget	**$197,700**	**Total Base Construction Budget**	**$209,800**
Replacement System Costs (HVAC and water heaters)	$9,000		
Total construction and replacement costs	**$206,700**	**Total construction costs**	**$209,800**
		Outdoor living with fireplace, summer kitchen	$9,500
		Rainwater: 5,000-gallon cistern	$4,500

Standard (Baseline) Home		Zero Cost Premium Home	
Description	Cost	Description	Cost
		6-kilowatt Solar PV system (after 30% federal tax credit)	$13,300
		Construction costs with amenities	$237,100
Site: suburban development on rural farmland	$46,500	Site: infill, city lot in high-density development with shared parking and community amenities	$143,300
Total land and construction costs	$253,200	Total land and construction costs	$380,400
Annual total cost of ownership (15-year amortization)	$16,880	Annual total cost of ownership (15-year amortization)	$25,360
Annual electric and water utility costs	$9,000	Annual electric and water costs from public utilities	$2,100
Average cleaning costs, including steam cleaning carpet	$1,500	Green cleaning products	$300
Annual maintenance costs	$2,500	Low-maintenance, durable materials, long life cycle performance, commissioning	$1,500
Life cycle annual redecorating costs	$1,500	No redecorating costs: premium paid initially for classic styles, materials, durable, natural materials	$0
From comparison in Chapter 1: annual property taxes and commute expenses	$9,900	Annual taxes and commute in urban high-density, mixed-use development, walkable distances and alternative transportation	$9,500
Total annual cost of ownership in traditional American home	$41,280	Total annual cost of ownership in zero cost premium home	$38,760
		Net annual savings to be applied to lifestyle or savings for retirement	$2,520

Zero Cost Premium

First, look how little more our base construction costs are for our zero cost premium home—$209,800 as compared to the traditional counterpart's $197,700. This provides definitive evidence that, using the strategies outlined in Section One of this book, we can build a healthy, high-performance home for not much more than our baseline home. This is due to the fact that we first focused on improving the efficiency of our home's design, and then treated the building as a system to specify size and improve each material and system in order to optimize total performance benefits. This allowed us to use fewer materials and smaller systems, saving money that paid for best practice materials and workmanship and high-performance system upgrades. In fact, an international study reported that green building only costs about a two percent premium over its traditionally built counterparts.[2]

Of course, some of those costs were based on choices that we made that you could easily change. First of all, the fees for our professional green architect reflect what we would pay for a custom home plan. This home was actually built by a small production builder, and he repeated the plan three more times in the same neighborhood. So, in reality, our actual amortized cost is one-quarter of the architect's fee, or $2,000.00, but we included the full $8,000 design fee in our calculations in our zero cost premium budget in order to prove that we could accomplish our goal even on a custom home project.

Note that we also paid significantly less for project management in our zero cost premium home than we did in the traditional home. This is because the builder of the traditional home has to pad his budget to cover materials misuse that leads not only to having to buy more materials but also increases construction waste disposal costs. Of course all builders should make a profit for their knowledge, skills and services, but it is not uncommon to see a 15–20 percent fee written into the contract. Our zero cost premium builder charged a flat fee, knowing that he was using highly skilled trades and that he would carefully manage materials use and therefore would not have any "surprise" cost overruns.

The detached carport saved materials and labor to the tune of $11,000 over the cost of an attached garage. We decided to upgrade from the 18-SEER (air conditioning efficiency) and 8.9-HSPF (heating efficiency) split system heat pump to ductless mini-splits in both of the condo units, due to their long life expectancy (25-year warranty). We could have stayed with the standard high-efficiency variable-speed split systems (two two-ton units) and saved several thousand dollars initially. However, a study[3] performed by the National Association of Home Builders and the Bank of America on the life expectancy of various home components indicates that we can only expect those units to last 10–15 years, so we then need to plan for the cost of replacement that would incur within the 15-year amortization period. Ditto for the standard water heaters, with an average life expectancy of only ten years. Just like with our metal roof, it turns out to be more cost effective to spend the additional funds up front to upgrade to solar thermal water heating and mini-split HVAC units. Although the initial expense is greater, the longer life expectancy and improved performance efficiencies make these investments worthwhile. This is especially true when you consider that you would have to replace both of these units within your initial 15-year amortization period, which would bring your total out-of-pocket costs to $206,700. That is within $3,100 of the total cost for our zero cost premium home without any of the affordable, healthy, high-performance benefits.

But our zero cost premium home is not complete until we add our rainwater collection system, outdoor living space and solar photovoltaic system. These are icing on the cake, so you could add them a few years after you have been in the home if they do not fit into your initial construction budget. It is important, though, that these are part of your long-term plan, as they are an integral component of our reduction in operational expenses that give us the desired reduced total cost of homeownership over time. Once we install and start using the rainwater and solar PV systems, our annual electric and water utility costs actually go to pay off those systems, so our net zero energy and reduced water pieces of the pie further speed up our payback.

Next, we recognize the long-term cost savings from site location. Note that both our projects are located in Austin, Texas, where land costs are much more expensive than most other areas of the country (and world!). In fact, you can buy a good infill lot in most other areas for about a third of what we have budgeted here. In other words, we've represented a worst-case scenario on land costs. In most other areas, the payback on our zero cost premium home would be even greater than what we indicate in this budget.

But what about that expensive infill lot? Many people would just walk away at that point, as they would not have the funds to invest in the site. But those costs may change, too. As urban densities increase, we will see fewer single-family homes occupying entire city lots. It is not uncommon now to see two homes sharing a lot, and in our case, that could be two stacked duplex units, effectively reducing the lot expense to each owner by 75 percent ($143,300/4 = $35,825 each), or much less than the $46,000 paid for the site in the suburban location.

Finally, remember that the construction costs for our zero cost premium home are for the single-family version of this home, the two-story version that you will raise your family in. Once the kids are grown and have moved out, you will spend a few thousand dollars adding the second-floor kitchen and the wall that closes off the stairwell to convert the house to a stacked duplex. At this point, you can recover half of your construction and land costs when you either sell the upper unit (and you downsize to the first-floor aging-in-place unit) or lease it for retirement income, whichever the case may be. At that point, the zero cost premium home budget wins hands-down over the traditional home, which would require a major expensive renovation to accomplish that same goal.

Now look at the amortized annual cost of ownership for the two budgets. Wow, we did it! Our net cost over the next 15 years would actually be less than what we would have expected to pay in higher utility costs, maintenance and repairs and commuting expenses over our baseline home. And that is at current rate structures for these costs. In reality we know that many unexpected events can and do occur that could cause these costs to be much more than what we have reflected in the model.

No cost premium for a healthy, high-performance green home! Not to mention the cost savings this represents to our community, our health and all the other stakeholders around the world who will benefit from our reduced impact on their cultures, ecosystems and quality of life. The overall potential cost savings from this approach to bring green mainstream is unimaginable and undeniable!

The Green Home Mortgage Industry

No discussion of the total-cost-of-ownership argument for the additional up-front costs of building a green home could be complete without a look at the financing aspects of those costs over time. In our comparison, the total construction and land costs for our zero cost premium home were $380,400 compared to $253,200 for the baseline home. Most of us don't have that kind of money lying around, so we must qualify for financing some part of it through a home mortgage.

Currently, the residential mortgage industry in the United States uses four criteria to qualify our ability to afford financing. This is typically referred to as PITI, or principal, interest, taxes and insurance. The sum of these annual costs must be equal to or less than some acceptable percentage of our annual income.

However, in light of the high number of bad mortgage loans and high foreclosure rates that led to the mortgage industry collapse and restructuring that has occurred over the last couple of years, more interest in encompassing the total cost of homeownership is being considered. As we discussed in Chapter 1, for instance, some of the highest foreclosure rates happened in bedroom communities created by sprawl, where rising property taxes and commuting expenses could not be sustained in the downturn of the economy.

New legislation[4] is being proposed that would base mortgage qualification on a broader base of related costs. This would be implemented in phases as methods for quantifying cost savings are developed in these related areas, including energy, water and lower commuting expenses. We think it would be great to analyze building components and methods for quality and durability as well. Including low maintenance, long life expectancy and lower repair and replacement costs over time should be something considered in mortgage qualifications. Third-party inspectors could verify compliance during construction.

It would be nice to think that, at some point, consideration of healthy home benefits would also be a consideration. Hopefully someone will start researching and documenting the healthcare cost savings of living in a green home with these features. But, as we stated previously, it is difficult to put a price on our health. But it's also difficult to think about losing your home to unexpected catastrophic healthcare costs. So, even if you never qualify for financing healthier building components, it's certainly worth your peace of mind to do so.

The Last Word: Go Do This!

It starts with you. And then you tell your friends, and they tell their friends. Builders will hear the market demand and flock to fill the niche. But, it does start with you.

To be certain that you find an architect and building team that understands green building, ask questions about the various aspects of design, site location and performance and make sure the answers align with what you have read here. Check references to verify that they are competent with the necessary experience to support your goals. If you hear from them or any of their project team that these methods won't work, fire them! An architect cannot argue that proper orientation, shading and simplistic design will not improve performance while lowering costs. A framer cannot argue that it takes less lumber in a home designed on two-foot modules, stacked stories, smaller footprint and using Advanced

Framing techniques. A plumber cannot argue that it takes less materials and labor to plumb a home with a central core plumbing design; ditto for the HVAC contractor with a centralized system location and a home designed for short, straight duct runs. And the same goes for the interior finish contractors (drywall, flooring, paint) and exterior finishes (masonry and roofing). Use your savings to invest in high-performance and healthy materials and systems and commissioning.

Find the right people that understand the benefits these strategies offer. Make sure everyone on the team is committed to working together to optimize performance and cost savings. Each experience moves more of these practices closer to the mainstream. As team members continue to develop their green expertise over time, it will eventually influence all of their continued work. Your efforts will contribute well beyond your own green home. You will have made a difference for all of us. Thank you and go forth and live well.

End Notes

Introduction

1. U.S. Department of Energy, *Buildings Energy Databook, 2006*, and *Annual Energy Review 2007*, June 2008, eia.doe.gov
2. U.S. Geological Survey, *Estimated Use of Water in the United States in 2005*, ga.water.usgs.gov/edu/wateruse/pdf/wutrends-2005.pdf
3. Nebraska Energy Office, *Minimizing the Use of Lumber Products in Residential Construction*, neo.ne.gov/home_const/factsheets/min_use_lumber, (accessed October 2013).
4. U.S. Environmental Protection Agency, *Estimating 2003 Building-Related Construction and Demolition Materials Amounts*, epa.gov/osw/conserve/imr/cdm/pubs/cd-meas.pdf (accessed October 2013).

Chapter 1: Location, Location, Location

1. United Nations Department of Economic and Social Affairs, *World Urbanization Prospects Highlights*, 2010, esa.un.org/unpd/wup/Documents/WUP2009_Highlights_Final.pdf (accessed November 2013).
2. Arthur C. Nelson, "Leadership in a New Era," *Journal of the American Planning Association*, Autumn, 2006, law.du.edu/images/uploads/rmlui/conferencematerials/2007/Thursday/DrNelsonLunchPresentation/NelsonJAPA2006.pdf
3. Rolf Pendall, Lesley Freiman, Dowell Myers and Selma Hepp, "Demographic Challenges and Opportunities for U.S. Housing Markets," bipartisanpolicy.org/library/report/demographic-challenges-and-opportunities-us-housing-markets (accessed March 2012).
4. smartgrowth.org
5. National Resources Conservation Service, "Wind Rose Data," wcc.nrcs.usda.gov/climate/windrose (accessed November 2013).
6. epa.gov/brownfields
7. epa.gov/superfund
8. Md. Wasim Siddiqui and R. S. Dhua, *Eating Artificially Ripened Fruits Is Harmful*, academia.edu/2321590/Eating_artificially_ripened_fruits_is_harmful (accessed November 2013).

Chapter 2: Size Matters

1. access-board.gov/guidelines-and-standards/buildings-and-sites

Chapter 3: Design, Design, Design

1. Joseph Lstiburek, *Roof Design*, buildingscience.com/documents /reports/rr-0404-roof-design (accessed October 2013).
2. U.S. Department of Energy, *Unvented Crawlspaces Code Adoption*, January, 2013, energy.gov/sites/prod/files/2014/01/f6/4_3c_ba_innov _unventedcrawlspaces_011713.pdf (accessed November 2013).
3. International Energy Conservation Code 2012, Sec. R 402.2.1
4. Alex Wilson, *Trombe Walls*, February 16, 2011, greenbuildingadvisor .com/blogs/dept/energy-solutions/trombe-walls (accessed October 2013).
5. R. K. Pletzer, J. W. Jones and B. D. Hunn, *Energy Savings Resulting from Shading Devices on Single-Family Residences in Austin, Texas*, Energy Systems Laboratory, Texas A&M University, repository.tamu.edu/ handle/1969.1/6484 (accessed October 2013).
6. Post-Tensioning Institute, *Standard Requirements for Design of Shallow Post-Tensioned Concrete Foundations on Expansive Soils*, Third Edition, May 2008. post-tensioning.org/Uploads/SOG/SOGDesign Std3rdEdv2008.pdf (accessed November 2013).
7. Itai Danielski, Morgan Fröling and Anna Joelsson, *The Impact of the Shape Factor on Final Energy Demand in Residential Buildings in Nordic Climates*, ases.conference-services.net/resources/252/2859/pdf /SOLAR2012_0428_full%20paper.pdf (accessed October 2013).
8. Rich Binsacca, *Flash Point*, February 10, 2011, builderonline.com /construction/flash-point (accessed October 2013).
9. Joseph Lstiburek and Terry Brennan, *Read This Before You Design, Build or Renovate*, apps1.eere.energy.gov/buildings/publications/pdfs /building_america/32114.pdf (accessed November 2013).
10. Kevin McCoy, *USA TODAY*, "Nation's Water Costs Rushing Higher," September 27, 2012, usatoday30.usatoday.com/money/economy /story/2012-09-27/water-rates-rising/57849626/1 (accessed October 2013).
11. EPA WaterSense, *Guide for Efficient Hot Water Delivery Systems*, August 30, 2012, epa.gov/watersense/docs/hw_distribution_guide.pdf (acessed November 2013).
12. Alliance for Water Efficiency, *Residential Hot Water Distribution Introduction*, allianceforwaterefficiency.org/Residential_Hot_Water _Distribution_System_Introduction (accessed November 2013).

13. Gary Klein, *Structured Plumbing Offers Real Benefits*, aim4sustainability.com/category/gary-klein-2 (accessed October 2013).

14. U.S. Environmental Protection Agency, *Right-Sized/Compact Ducts*, energystar.gov/ia/home_improvement/home_sealing/RightSized _CompactDuctsFS_2005.pdf (accessed October 2013).

15. Doug Garrett, *Home Energy Magazine*, "Beware the Closed Bedroom Door," January 1, 2001, homeenergy.org/show/article/id/1704 (acessed October 2013).

16. Center for Power Electronics Systems, *Future Home DC-based Renewable Energy nanoGrid System*, cpes.vt.edu/public/showcase/d1.2 FutureHome (accessed October 2013).

17. Penn State University, *Room Calculations*, personal.psu.edu/users/a/s /asp5045/roomcalclivingroom.htm (accessed October 2013).

18. *Inside the Designer's Studio 37*, hoklife.com/2010/04/06/inside-the -designers-studio-37hok-ceo-patrick-macleamy (accessed October 2013).

19. Andrea Korber and Brad Guy, *Design for Disassembly in the Built Environment: An Atlanta Home Case Study*, U.S. Environmental Protection Agency, lifecyclebuilding.org/docs/DfDCaseStudyHome Summary.pdf

20. Eric Corey Freed and Kevin Daum. *Green$ense for the Home: Rating the Real Payoff from 50 Green Home Projects*, Taunton, 2010. p. 278.

Chapter 4: Building Materials: Shades of Green

1. TerraChoice Environmental Marketing, *Greenwashing Affects 98% Of Products Including Toys, Baby Products and Cosmetics*, April 15, 2009, terrachoice.com/images/Seven%20Sins%20of%20Greenwashing%20 Release%20-%20April%2015%202009%20-%20US.pdf (accessed November 2013).

2. William McDonough and Michael Braungart. *Cradle to Cradle: Remaking the Way We Make Things*, North Point, 2002. p. 27.

3. Global Footprint Network, *Living Planet Report 2006*," footprintnet work.org/download.php?id=308 (accessed November 2013).

4. Rainforest Relief, *Avoiding Unsustainable Rainforest Wood*, rainforest relief.org/What_to_Avoid_and_Alternatives/Rainforest_Wood, (accessed November 2013).

5. jameshardie.com/homeowner/products-colorplus

6. Jan Kośny, PhD, David W. Yarbrough, PhD, PE, and Phillip Childs. "Couple Secrets about How Framing Is Effecting the Thermal Performance of Wood and Steel-Framed Walls." Oak Ridge National

Laboratory, 15 Sept. 2006. Web. web.ornl.gov/sci/roofs+walls/research /detailed_papers/thermal_frame/

7. Eric Corey Freed and Kevin Daum, *Greensense for the Home: Rating the Real Payoff from 50 Green Home Projects*, Taunton Press, 2010. p. 245.

8. *Wood Truss & Wall Panels*, classictruss.com/structural_components _vs_stick_framing.asp (accessed November 2013).

9. J. Kośny, J. E. Christian and A. O. Desjarlais, *Metal Stud Wall Systems—Thermal Disaster, or Modern Wall Systems With Highly Efficient Thermal Insulation?* Oak Ridge National Laboratory, Buildings Technology Center, astm.org/DIGITAL_LIBRARY/STP/PAGES/STP 12273S (accessed November 2013).

10. Robert L. Roy, *Earth-sheltered Houses: How to Build an Affordable Underground Home*, New Society Pulishers, 2006, p. 6.

11. Marc Lalanillia, *The Benefits of IPM*, greenliving.about.com/od/green livingbasics/a/ipm (accessed December 2013).

12. U.S. Environtmental Protection Agency, *Vacate and Safe Re-Entry Time*, epa.gov/dfe/pubs/projects/spf/when_is_it_safe_to_re-enter _after_spf_installation (accessed November 2013).

13. Jeffrey Christian and Jan Kosny, *Home Energy Magazine Onlne*, "Calculating Whole Wall R-Values on the Net," homeenergy.org/show /article/id/1517 (accessed November 2013).

14. Allison Bailes, *Energy Vanguard Blog*, "ENERGY STAR Version 3 Requires Grade I Insulation," energyvanguard.com/blog-building -science-HERS-BPI/bid/36110/ENERGY-STAR-Version-3-Requires -Grade-I-Insulation (accessed December 2013).

15. ENERGY STAR, *High Performance Windows*, energystar.gov/index .cfm?c=new_homes_features.hm_f_advanced_windows (accessed November 2013).

16. National Fenestration Rating Council, *The NFRC Label*, nfrc.org /WindowRatings/The-NFRC-Label (accessed November 2013).

17. Heat Island Group, *Cool Science: Cool Roofs*, heatisland.lbl.gov/cool science/cool-science-cool-roofs (accessed November 2013).

18. William (Bill) Miller, PhD, Hashem Akbari, PhD, et. al., "Special Infrared Reflective Pigments Make a Dark Roof Reflect Almost Like a White Roof," web.ornl.gov/sci/roofs+walls/staff/papers/new_53.pdf (accessed November 2013).

19. Florida Division of Emergency Management, Hurricane Retrofit Guide, *Wood Frame Roof-to-Wall Connections*, floridadisaster.org /hrg/content/walls/wood_frame_rtw_conn (accessed November 2013).

20. U.S. Department of Energy, *Advanced Wall Framing*, apps1.eere .energy.gov/buildings/publications/pdfs/building_america/26449.pdf (accessed October 2013).

21. Joseph Lstiburek, *Building Science Insights: Advanced Framing*, Building Science Corporation, February 2010, buildingscience.com/docu ments/insights/bsi-030-advanced-framing (accessed November 2013).

22. Ibid.

23 Joseph Lstiburek, *Basement Insulation Systems*, apps1.eere.energy.gov /buildings/publications/pdfs/building_america/1_1a_ba_innov_base mentinsulationsystems_011713.pdf (accessed December 2013).

24. U.S. Department of Energy, *Thermal Bypass Checklist*, energystar.gov /ia/partners/bldrs_lenders_raters/downloads/Thermal_Bypass_Ins pection_Checklist.pdf (accessed November 2013).

25. *ASTM E283 - 04(2012) Standard Test Method for Determining Rate of Air Leakage Through Exterior Windows, Curtain Walls, and Doors Under Specified Pressure Differences Across the Specimen*, astm.org /Standards/E283 (accessed November 2013).

26. Ingrid Melody, *Radiant Barriers: A Question & Answer Primer*, Florida Solar Energy Center, fsec.ucf.edu/en/publications/pdf/FSEC-EN -15-87.pdf (accessed October 2013).

27. Joseph Lstiburek, *Water-Managed Wall Systems*, buildingscience.com /documents/published-articles/pa-water-managed-wall-systems) accessed November 2013).

28. *2012 International Building Code*, Section 2510.6, publicecodes.cyber regs.com/icod/ibc/2012/icod_ibc_2012_25_par042 (accessed March 2014).

29. Building Science Corporation, *Basement Insulation Systems*, 2002, buildingscience.com/documents/bareports/ba-0202-basement-insul ation-systems/view?searchterm=basement%20insulation%20systems (accessed November 2013).

30. Martin Holladay, *How to Insulate a Basement Wall*," June 29, 2012, greenbuildingadvisor.com/blogs/dept/musings/how-insulate-base ment-wall (accessed November 2013).

31. Joseph Lstiburek, *Basement Insulation Systems*, U.S. Department of Energy, January 2013, apps1.eere.energy.gov/buildings/publications /pdfs/building_america/1_1a_ba_innov_basementinsulationsystems _011713.pdf (accessed December 2013).

32. Joseph Lstiburek, *Understanding Basements*, Building Science Corporation, October 27, 2006, buildingscience.com/documents/digests /bsd-103-understanding-basements (accessed November 2013).

33. Ted Rieger, *Drawbacks of Powered Attic Ventilators,"* Home Energy, November–December, 1995, homeenergy.org/show/article/year/1995 /id/1165 (accessed December 2013).

34. Craig DeWitt, *The Fallacies of Venting Crawl Spaces,* January 3, 2002, rlcengineering.com/csfallacies (accessed December 2013).

35. U.S. Department of Energy, Building Technologies program, *Unvented, Conditioned Crawlspaces,* January 2013, apps1.eere.energy.gov /buildings/publications/pdfs/building_america/1_1d_ba_innov_un ventedconditionedcrawlspaces_011713.pdf (accessed December 2013).

36. Joseph Lstiburek, Building Science Corporation, *Building Science Insights: The Perfect Wall,* May, 2008, buildingscience.com/documents /insights/bsi-001-the-perfect-wall (accessed December 2013).

37. Ibid.

Chapter 5: Construction Waste

1. Home Innovation Research Labs, *Residential Construction Waste: From Disposal to Management,* 2001, toolbase.org/Best-Practices /Construction-Waste/residential-construction-waste (accessed October 2013).

2. Ann Virginia Edminster, Semi Yassa and Matthew McDermid, *Efficient Wood Use in Residential Construction: A Practical Guide to Saving Wood, Money and Forests,* Natural Resources Defense Council, 1998, nrdc.org/cities/building/rwoodus.asp (accessed November 2013).

3. CE Green Building Resource Center, *Construction and Demolition Waste,* codegreenhouston.org/materials-main-page/construction -and-demolition-waste (accessed November 2013).

4. J. M. Sillick and W. R. Jacobi, *Healthy Roots and Healthy Trees,* Colorado State University Extension, March 2009, ext.colostate.edu/pubs /garden/02926 (accessed November 2013).

5. Home Innovation Research Labs, *Residential Construction Waste: From Disposal to Management,* 2001, toolbase.org/Best-Practices /Construction-Waste/residential-construction-waste (accessed October 2013).

6. U.S. Department of Energy, National Renewable Energy Laboratory, *Technology Fact Sheet: Advanced Wall Framing,* October 2000, apps1 .eere.energy.gov/buildings/publications/pdfs/building_america/264 49.pdf (accessed November 2013).

Chapter 6: Equipment and Systems

1. U.S. EPA Energy Star Program, *Right-Sized Air Conditioners: Mechanical Equipment Improvements,"* energystar.gov/ia/home_improvement

/home_sealing/RightSized_AirCondFS_2005.pdf (accessed November 2013).

2. U.S. Department of Energy, *Minimize Boiler Short Cycling Losses*, eere.energy.gov/industry/.../pdfs/steam16_cycling_losses.pdf (accessed November 2013).

3. Martin Holladay, "Saving Energy With Manual J and Manual D," *Green Building Advisor*, August 13, 2010, greenbuildingadvisor.com/blogs/dept/musings/saving-energy-manual-j-and-manual-d (accessed November 2013).

4. Rob Falke, "Don't Exceed Your Static Pressure Budgets Part 1," *Contracting Business*, September 1 2007, contractingbusiness.com/enews letters/cb_imp_72292 (accessed November 2013).

5. D. S. Parker, J. R. Sherwin, R. A. Raustad and D. B. Shirey III, *Impact of Evaporator Coil Air Flow in Residential Air Conditioning Systems*, Florida Solar Energy Center, fsec.ucf.edu/en/publications/html /FSEC-PF-321-97/index (accessed October 2013).

6. Allison A. Bailes III, PhD, "How to Tell If You Have an Oversized Air Conditioner," *Energy Vanguard Blog*, June 14 2010, energyvanguard .com/blog-building-science-HERS-BPI/bid/24645/How-to-Tell-If -You-Have-an-Oversized-Air-Conditioner (accessed November 2013).

7. U.S. EPA Energy Star Program, *Right-Sized Air Conditioners Mechanical Equipment Improvements*, energystar.gov/ia/home_improvement /home_sealing/RightSized_AirCondFS_2005.pdf (accessed November 2013).

8. U.S. Department of Energy, "Electric Resistance Heating," *Energy. gov*, June 24 2012, energy.gov/energysaver/articles/electric-resistance -heating (accessed November 2013).

9. U.S. Department of Energy, "Geothermal Heat Pumps," *Whole Building Design Guide*, August 24 2012, wbdg.org/resources/geothermal heatpumps.php (accessed November 2013).

10. U.S. Department of Energy, "Evaporative Coolers," *Energy.gov*, July 1 2012, energy.gov/energysaver/articles/evaporative-coolers (accessed November 2013).

11. Home Innovation Research Labs, *Combined Heat and Power Systems for Residential Use*, 2001, toolbase.org/Technology-Inventory/Electri cal-Electronics/combined-heat-power (accessed November 2013).

12. U.S. Energy Information Administration, *Heating Fuel Comparison Calculator*, eia.gov/neic/experts/heatcalc.xls (accessed November 2013).

13. U.S. Energy Information Administration, *Total Energy: Monthly Energy Review*, eia.gov/totalenergy/data/monthly/#price (accessed November 2013).

14. U.S. Department of Energy, "Electric Resistance Heating," *Energy.gov*, June 24 2012, energy.gov/energysaver/articles/electric-resistance-heating (accessed November 2013).

15. Worldwatch Institute, *Real Organic Agriculture: Using Human Waste as Fertilizer*, 2013, worldwatch.org/node/5394 (accessed November 2013).

16. Home Innovation Research Labs, *Information-Age Wiring for Home Automation Systems*, toolbase.org/Technology-Inventory/Electrical-Electronics/home-automation-wiring (accessed November 2013).

17. International Energy Conservation Code 2012, Sec. R 404.1

18. U.S. Department of Energy, *DSIRE: Database of Energy Efficiency, Renewable Energy Solar Incentives, Rebates, Programs, Policy*, dsireusa.org (accessed November 2013).

19. Natural Resources Canada, *Solar Ready Guidelines for Solar Domestic Hot Water and Photovoltaic Systems*, nrcan.gc.ca/energy/publications/sciences-technology/housing/6295 (accessed November 2013).

20. U.S. Department of Energy, Energy Efficiency & Renewable Energy, *A Homebuilder's Guide to Going Solar*, April 2008, adeca.alabama.gov/Divisions/energy/Documents/Consumer%20docs/EERE-Home%20Builders%20Guide%20to%20Going%20Solar.pdf (accessed November 2013).

Chapter 7: Health

1. GoodGuide, *The Importance of Basic Toxicity Testing*, 2011, scorecard.goodguide.com/chemical-profiles/def/basic_det (accessed November 2013).

2. Safer Chemicals, Healthy Families, *"Safe Chemicals Act of 2011" Introduced Today Legislation Would Protect American Families from Toxic Chemicals*, April 14 2011. saferchemicals.org/2011/04/safe-chemicals-act-of-2011-introduced-today-legislation-would-protect-american-families-from-toxic-chemicals (accessed November 2013).

3. U.S. Environmental Protection Agency, *HPV Chemical Hazard Data Availability Study*, epa.gov/hpv/pubs/general/hazchem (accessed November 2013).

4. U.S. Environmental Protection Agency, *Volatile Organic Compounds (VOCs)*, November 8 2012., epa.gov/iaq/voc2 (accessed November 2013).

5. U.S. Centers for Disease Control and Prevention, *Asthma in the US: Growing Every Year*, May 2011, cdc.gov/VitalSigns/Asthma/ (accessed November 2013).

6. U.S. Centers for Disease Control and Prevention, *Why Are Autism*

Spectrum Disorders Increasing? April 2012, cdc.gov/Features/Autism Prevalence (accessed November 2013).

7. Wayne R. Ott and John W. Roberts, *Everyday Exposure to Toxic Pollutants*, February 1998, 0goapes.weebly.com/uploads/3/2/3/9/3239894 /everyday_exposure_to_toxic_chemicals.pdf (accessed November 2013).

8. National Toxicology Program, *12th Report on Carcinogens (RoC)*, June 2011, ntp.niehs.nih.gov/?objectid=03C9AF75-E1BF-FF40-DBA9EC09 28DF8B15 (accessed November 2013).

9. Healthy Building Network, Kaiser Permanente, *Toxic Chemicals in Building Materials*, May 2008. healthybuilding.net/healthcare/Toxic %20Chemicals%20in%20Building%20Materials.pdf (accessed November 2013).

10. U.S. Environmental Protection Agency, *Indoor Air Package Specifications, Version 2*, Sections 5.5, 5.6, 5.7, April 2007. energystar.gov/ia /.../IAP_Specification_041907.pdf (accessed December 2013).

11. Puget Sound Clean Energy Agency, *Facts About Burning Wood*, November 2002, docstoc.com/docs/2299283/Facts-about-burning-wood (accessed October 2013).

12. Dr. Wayne Ott, quoted in *Burning Issues Wood Smoke Fact Sheet*, February 1 1998, burningissues.org/fact-sheet.htm (accessed November 2013).

13. Nick Gromicko, *Ventless Fireplace Inspection*, nachi.org/ventless-fire place-inspection (accessed November 2013).

14. U.S. Environmental Protection Agency, *Air Pollution Control Technology Fact Sheet*, epa.gov/ttncatc1/dir1/fsetling.pdf (accessed November 2013).

15. Phil Lieberman, MD, American College of Allergy, Asthma and Immunology, *The Relationship Between House Dust Mite Growth and Relative Humidity*, May 2 2012. aaaai.org/ask-the-expert/dust-mite -growth-relative-humidity.aspx (accessed November 2013).

16. U.S. Environmental Protection Agency, *Mold Resources*, epa.gov/mold /moldresources.html (accessed November 18 2013).

17. U.S. Department of Energy, ENERGY STAR *Qualified Products*, energy star.gov/index.cfm?fuseaction=find_a_product. (accessed November 2013).

18. Green Seal, *Green Seal Standard for Paints and Coatings*, January 2010, greenseal.org/Portals/0/Documents/Standards/GS-11/GS-11_Ed3-1 _Paints_and_Coatings.pdf (accessed November 2013).

19. Healthy House Institute, *The EMF Controversy*, healthyhouseinstitute .com/a-721-The-EMF-Controversy (accessed November 2013).

20. International Organization for Standardization, *ISO 14000—Environmental Management*, iso.org/iso/iso14000 (accessed November 2013).

21. Blair E. Witherington, *Behavioral Responses of Nesting Sea Turtles to Artificial Light*, 1992, jstor.org/discover/10.2307/3892916?uid=3739920&uid=2&uid=4&uid=3739256&sid=21103388181257 (accessed November 2013).

22. Travis Longcore and Catherine Rich, *Light Pollution and Ecosystems*, American Institute of Biological Studies, May 2010, actionbioscience.org/environment/longcore_rich (accessed November 2013).

23. Patricia A. Niquette AuD, *Noise Exposure: Explanation of OSHA and NIOSH SafeExposure Limits and the Importance of Noise Dosimetry*, Etymotic Research, etymotic.com/pdf/er_noise_exposure_white paper.pdf (accessed November 2013).

Chapter 8:Outdoor Living

1. Howard Garrett, *Weed & Feed Fertilizers*, The Dirt Doctor, 2013, dirtdoctor.com/Weed-Feed-Fertilizers_vq1334 (accessed November 2013).

2. "Austin Guide to...Avoiding Weed and Feed," austintexas.gov/sites/default/files/files/Watershed/growgreen/weedandfeed.pdf (accessed November 2013).

3. National Wildlife federation, *Wildlife Habitat Certification*, nwf.org/CertifiedWildlifeHabitat (accessed November 2013).

Chapter 9: Green Bling

1. U.S. Department of Energy, *Recommended Levels of Insulation*, energy star.gov/index.cfm?c=home_sealing.hm_improvement_insulation_table (accessed November 2013).

Chapter 10: Keeping It Green

1. Lynn Knight, Arlene Levin and Katherine Mendenhall, U.S. Environmental Protection Agency, *Candles and Incense as Potential Sources of Indoor Air Pollution*, January 1 2001, nepis.epa.gov/Adobe/PDF/P1009D5G.pdf (accessed November 2013).

2. Healthy Child Healthy World, *A Wake-Up Story*, healthychild.org/healthy-living/a-wake-up-story (accessed November 2013).

3. National Park Service, *Green Power*, November 7 2013, nps.gov/climatefriendlyparks/involved/resources/greenpower (accessed November 2013).

4. U.S. Environmental Protection Agency, *Renewable Energy Credits (RECs)*, epa.gov/greenpower/gpmarket/rec (accessed November 2013).

Chapter 11: The Net Zero Energy Capable Home Model

1. U.S Census Bureau, census.gov/const/C25Ann/sftotalmedavgsqft.pdf (accessed November 2013).
2. Department of Energy, *Home Energy Saver*, http://hes.lbl.gov /consumer (accessed November 2013).
3. Sarah Susanka and Kira Obolensky, *The Not So Big House*, Taunton, 1998.
4. John Proctor, Brad Wilson and Zinoviy Katsnelson, "Bigger Is Not Better: Sizing Air Conditioners Properly," *Home Energy Magazine*, May–June 1995, homeenergy.org/show/article/year/1995/magazine /91/id/1128 (accessed November 2013). 2013.
5. Ibid.
6. Air Conditioning Contractors of America, *Quality Standards*, acca .org/standards/quality (accessed November 2013).
7. Martin Holladay, "HRV or ERV?" *Green Building Advisor*, greenbuild ingadvisor.com/blogs/dept/musings/hrv-or-erv (accessed November 2013).
8. Oikos Green Building Library, *Water Heaters & Energy Efficiency*, oikos.com/library/energy_outlet/water_heaters (accessed March 2014).
9. Efficiency Vermont, *ENERGY STAR Most Efficient and CEE Tiers 2 & 3 Refrigerators Eligible for Efficiency Vermont Rebate*, efficiency vermont.com/docs/for_my_home/rebate_forms/EVT_QPL_Refriger ators.pdf (accessed October 2013).
10. Home Innovation Research Labs, *Induction Cooktops*, toolbase.org /Technology-Inventory/Appliances/induction-cooktops (accessed March 2014).
11. Karen Ehrhardt-Martinez, Kat A. Donnelly and John A. "Skip" Laitner, *Advanced Metering Initiatives and Residential Feedback Programs: A Meta-Review for Household Electricity-Saving Opportunities*, American Council for an Energy-Efficient Economy, June 2010, smartgrid .gov/sites/default/files/pdfs/ami_initiatives_aceee.pdf (accessed November 2013).

Chapter 12: Net Zero Water

1. J. Alex Forasté, P. E. and David Hirschman, *A Methodology for Using Rainwater Harvesting as a Stormwater Management BMP*, lshs.tamu .edu/docs/lshs/end-notes/a%20methodology%20for%20using%20

rainwater%20harvesting-4168265292/a%20methodology%20for%20 using%20rainwater%20harvesting.pdf (accessed November 2013).

2. U.S. Environmental Protection Agency, *Water Facts*, June 2004, water.epa.gov/lawsregs/guidance/sdwa/upload/2009_08_28_sdwa _fs_30ann_waterfacts_web.pdf (accessed November 2013).

3. John J. Sloan, Peter A.Y. Ampim, Raul I. Cabrera, Wayne A. Mackay and Steve W. George, *Moisture and Nutrient Storage Capacity of Calcined Expanded Shale, Principles, Application and Assessment in Soil Science*, cdn.intechopen.com/pdfs/24767/InTech-Moisture_and _nutrient_storage_capacity_of_calcined_expanded_shale.pdf. p. 35. (accessed November 2013).

4. Environmental Protection Agency, *Tips for Watering Wisely: Sprinkler Spruce-Up*, epa.gov/watersense/outdoor/watering_tips (accessed November 2013).

5. Alliance for Water Efficiency, *Residential End Uses of Water Study (1999)*, allianceforwaterefficiency.org/residential-end-uses-of-water -study-1999.aspx (accessed November 2013).

6. U.S. Environmental Protection Agency, *Exercise II. The Superior Car Wash*, water.epa.gov/learn/resources/midsh/stopx2.cfm (accessed November 2013).

7. Bahman Sheikh, PhD, PE, *White Paper on Graywater*, 2010, graywater .org.il/Documents/GraywaterFinal%20Report2010%20for%20the%20 US%20-%20AWWA.pdf (accessed November 2013).

8. Alex Wilson and Rachel Navaro, *Environmental Building News*, "Alternative Water Sources: Supply-Side Solutions for Green Buildings," buildinggreen.com/auth/article.cfm/2008/4/29/Alternative-Water -Sources-Supply-Side-Solutions-for-Green-Buildings (accessed November 2013).

9. Nicki Meier, *Water Retention Landscape Techniques for Farm and Garden*, The Permaculture Research Institute, August 8 2013, perma culturenews.org/2013/08/08/water-retention-landscape-techniques -for-farm-and-garden (accessed November 2013).

10. Dayna Yocum, *Design Manual: Greywater Biofiltration, Constructed Wetland System*, fiesta.bren.ucsb.edu/~chiapas2/Water%20Manage ment_files/Greywater%20Wetlands-1.pdf (accessed November 2013).

11. Tara Franey, *Composting Toilets: Alleviating Regulatory Barriers to an Integrated Green Solution*, vermontlaw.edu/Documents/Land%20 Use%20Institute/NEWEA_paper_final_1.pdf (accessed November 2013).

12. U.S. Geological Survey, *The World's Water.* ga.water.usgs.gov/edu /earthwherewater (accessed November 2013).

13. Jeremy Hance, *Earth's Ecosystems Still Soaking up Half of Human Carbon Emissions*, Mongabay.com. August 6 2012, news.mongabay.com /2012/0806-hance-earth-sequestration (accessed November 2013).

14. Worldwatch Institute, *Oceans in Peril: Protecting Marine Biodiversity*, September 2007, worldwatch.org/node/5352 (accessed November 2013).

15. Lone Star Chapter of the Sierra Club, *Desalination: Is It Worth Its Salt?* November 2011, texas.sierraclub.org/press/Desalination.pdf (accessed November 2013).

Chapter 13: Zero Waste

1. Connecticut Department of Energy & Environmental Protection, *Reduce Reuse Recycle*, August 28 2013, ct.gov/deep/cwp/browse.asp?A =2714 (accessed November 2013).

2. U.S. Environmental Protection Agency, *Reducing Food Waste for Businesses*, epa.gov/foodrecovery/ (accessed November 2013).

3. U.S. Environmental Protection Agency, *Waste Prevention*, epa.gov /wastes/conserve/smm/wastewise/wrr/prevent (accessed November 2013).

4. Wisconsin Department of Natural Resources, *Recycling and Waste Reduction: A Guide for the Workplace*, dnr.wi.gov/files/PDF/pubs/wa /WA1533.pdf (accessed November 2013).

5. U.S. Environmental Protection Agency, *Environmental Management Systems*, epa.gov/wastes/conserve/smm/wastewise/wrr/ems (accessed November 2013).

6. U.S. Environmental Protection Agency, *History of RCRA*, epa.gov /waste/laws-regs/rcrahistory.htm (accessed November 2013).

7. U.S. Environmental Protection Agency, *Green Building: Reducing Waste*, epa.gov/greenhomes/ReduceWaste (accessed November 2013).

8. VIRIDIS, Wentworth Design Green, *Engineered Wood: Re-Use and Recycling*, myweb.wit.edu/viridis/green_site/projects/1_materials /engineered_wood/4_reuse/reuse (accessed November 2013).

9. Annie Leonard, Story of Stuff Project, *Story of Change*, storyofstuff .org/movies/story-of-change (accessed November 2013).

10. Pennsylvania Department of Environmental Protection, *Waste Reduction in the Home*, dep.state.pa.us/dep/deputate/airwaste/wm /recycle/FACTS/Reduce (accessed November 2013).

11. Landfill Reduction & Recycling, Inc., *Landfill Reduction & Recycling*, landfillreduction.com/Home_Page (accessed November 2013).

12. Providence Journal, *Landfill Operator Distributes $1.9 Million in Recycling Profits*, September 6 2012, news.providencejournal.com/break ing-news/2012/09/johnston-ri----5 (accessed November 2013).

13. Whatcom County Agriculture, *Composting with Redworms*, whatcom .wsu.edu/ag/compost/Redwormsedit (accessed November 2013).

14. Zenhabits, *The Habit Change Cheatsheet: 29 Ways to Successfully Ingrain a Behavior*, September 29 2009, zenhabits.net/the-habit-change -cheatsheet-29-ways-to-successfully-ingrain-a-behavior (accessed November 2013).

Chapter 14: Zero Your Carbon Footprint

1. Intergovernmental Panel on Climate Change, *Summary for Policy Makers*, September 27 2013, ipcc.ch (accessed November 2013).

2. T. Ramesh, R. Prakash and K.K. Shukla, *Energy and Buildings*, "Life Cycle Energy Analysis of Buildings: An Overview," volume 42, pp. 1592–1600, 2010.

3. U.S. Environmental Protection Agency, *Our Built and Natural Environments: A Technical Review of the Interactions Between Land Use, Transportation, and Environmental Quality (2nd Edition)*, epa.gov /smartgrowth/built (accessed November 2013).

4. Roger S. Swift, "Sequestration of Carbon by Soil," *Soil Science*, 166.11, 2001, pp. 858–871.

5. An Teach Glas: High Performance Zero Carbon Passive House, *Low Carbon Construction*, zerocarbonpassivehouse.com/loe-carbon -construction (accessed November 2013).

6. U.S. Environmental Protection Agency, *Overview of Greenhouse Gases: Methane Emissions*, September 9 2013, epa.gov/climatechange /ghgemissions/gases/ch4 (accessed November 2013).

7. Washington State University Extension Energy Program, *Variable Refrigerant Flow (VRF) Heat Pumps*, e3tnw.org/ItemDetail.aspx?id=200 (accessed November 2013).

8. Scottsdale Green Building Program, *Building Materials*, scottsdaleaz .gov/Assets/Public%20Website/greenbuilding/GBMaterials.pdf (accessed November 2013).

9. U.S. Environmental Protection Agency, *Green Roofs*, epa.gov/hiri /mitigation/greenroofs (accessed November 2013).

10. Carbon Footprint, *Carbon Footprint Calculator*, carbonfootprint.com /calculator.aspx (accessed November 2013).

Chapter 15: Zero Cost Premium

1. Rolf Pendall, Lesley Freiman, Dowell Myers and Selma Hepp, Bipartisan Policy Center, *Demographic Challenges and Opportunities for U.S. Housing Markets*, bipartisanpolicy.org/library/report/demographic

-challenges-and-opportunities-us-housing-markets (accessed November 2013).

2. Greg Kats, Good Energies, *Greening Buildings and Communities: Costs and Benefits*, goodenergies.com/news/-pdfs/Web%20site%20 Presentation.pdf (accessed November 2013).

3. National Association of Home Builders and Bank of America Home Equity, *Study of Life Expectancy of Home Components*, February 2007, nahb.org/fileUpload_details.aspx?contentID=99359 (accessed November 2013).

4. Michael Bennett and Johnny Isakson. *S. 1106*, gpo.gov/fdsys/pkg /BILLS-113s1106is/pdf/BILLS-113s1106is.pdf (accessed November 2013).

Index

ACCA Quality Installation procedure, 296

accessibility, 34–35

acoustical properties, 111–12, 115, 126

adhesives, 127, 220, 231, 232, 369

Advanced Framing, 66, 67, 128–32, 160–61, 173

aesthetics, 103, 308

agriboard, 115, 127

air barriers, 81, 135–38, 139, 143, 149, 293

air circulation, 49–55. *See also* fresh air ventilation systems.

air conditioner equipment, 44, 45, 176. *See also* condensate; HVAC.

Air Conditioning Contractors of America (ACCA), 80, 179

air handlers, 80, 83, 160, 260. *See also* furnaces; HVAC.

air leakage (AL), 120, 121

air quality. *See* indoor air quality; outdoor air quality.

allergies, 217, 218–19, 270, 271

alternating current (AC), 85

alternative building materials, 115, 135

Americans with Disabilities Act, 34–35

American Water Works Association Research Foundation (AWWARF), 313

annual energy consumption, 289, 293, 298, 300

Annual Fuel Utilization Efficiency (AFUE), 183

apartment, in a home, 33, 34, 36. *See also* convertible house.

appliances, 206–07, 260, 272–73, 351. *See also by name.*

in net zero home model, 297–98, 313–15

arbors, 111, 275, 277

architects, 160, 191–92, 214

AR4 Synthesis Report, 343

artificial turf, 309

ASHRAE standards, 231, 287, 368

asphalt shingles, 107, 123, 170

ASTM E-283, 140

Athena Sustainable Materials Institute, 361

Atrazine, 248

attics, 44–46, 47, 81, 118, 134, 137–38

autoclaved aerated concrete blocks (AACs), 115, 135

automobile use, 345, 355–56

awning structures, 57–58

baby boomers, 12, 25, 376

bamboo, 108, 126, 349

base construction costs, 31, 66, 256, 383

baseline energy model home, 285–89
building specifications, 287–89
design, 285–87

basements, 41, 110–11, 135, 146–49, 228–29

BASF, 372–75

bathrooms, 74, 75, 88, 92, 201, 230–31, 261, 297

battery storage, 211

batt insulation. *See* fiberglass insulation.

bedrooms, 61, 63, 64

berms, 247, 324

Bernoulli principle, 52

About the Authors

MIKI COOK is a green building and sustainability consultant for the oldest green building program in the nation, Austin Energy Green Building. She has spent her career in the residential construction industry, where she gained a thorough understanding of green strategies and costs. Miki has also served as a Green Rater for the U.S. Green Building Council's LEED for Homes, Energy Star, and ICC-700, the national green building standard. More recently, Miki has dedicated her career to educating contractors and the public on the strategies, methods and benefits of green homes, and helping them to achieve those goals.

DOUG GARRETT, CEM, founded Austin Energy's Residential and Multi-Family Energy Conservation programs. He has studied building science, indoor air quality, moisture management, and heating, ventilation and air conditioning throughout his career. In 1996, Doug established the first building science consulting business in Texas and continues to provide building science-based forensic investigations, diagnostics and design consultations for clients and homebuilders across the nation. Doug has presented hundreds of seminars on applied building science, energy efficiency, moisture management, energy codes, air conditioning, indoor air quality and green building. He has served on numerous advisory councils and task forces for high-performance industry standards, energy codes, alternative fuels, health concerns and housing affordability.

If you have enjoyed *Green Home Building* you might also enjoy other

BOOKS TO BUILD A NEW SOCIETY

Our books provide positive solutions for people who want to make a difference. We specialize in:

Sustainable Living • Green Building • Peak Oil •
Renewable Energy • Environment & Economy Natural
Building & Appropriate Technology • Progressive Leadership
Resistance and Community • Educational & Parenting Resources

New Society Publishers

ENVIRONMENTAL BENEFITS STATEMENT

New Society Publishers has chosen to produce this book on recycled paper made with **100% post consumer waste,** processed chlorine free, and old growth free.

For every 5,000 books printed, New Society saves the following resources:[1]

44	Trees
3,988	Pounds of Solid Waste
4,388	Gallons of Water
5,724	Kilowatt Hours of Electricity
7,250	Pounds of Greenhouse Gases
31	Pounds of HAPs, VOCs, and AOX Combined
11	Cubic Yards of Landfill Space

[1]Environmental benefits are calculated based on research done by the Environmental Defense Fund and other members of the Paper Task Force who study the environmental impacts of the paper industry.

For a full list of NSP's titles, please call 1-800-567-6772 *or visit our website* at:

www.newsociety.com